Structured Creativity

Also by Hugo Tschirky

TECHNOLOGY AND INNOVATION MANAGEMENT ON THE MOVE:
From Managing Technologies to Managing Innovation-driven Enterprises
(*with P. Savioz and H.-H. Jung*)

BRINGING TECHNOLOGY INTO THE BOARDROOM (*co-author*)

Structured Creativity

Formulating an Innovation Strategy

Tim Sauber

and

Hugo Tschirky

in association with the
EUROPEAN INSTITUTE FOR TECHNOLOGY
AND INNOVATION MANAGEMENT

© Tim Sauber 2006

All rights reserved. No reproduction, copy or transmission of this publication may be made without written permission.

No paragraph of this publication may be reproduced, copied or transmitted save with written permission or in accordance with the provisions of the Copyright, Designs and Patents Act 1988, or under the terms of any licence permitting limited copying issued by the Copyright Licensing Agency, 90 Tottenham Court Road, London W1T 4LP.

Any person who does any unauthorized act in relation to this publication may be liable to criminal prosecution and civil claims for damages.

The authors have asserted their rights to be identified as the authors of this work in accordance with the Copyright, Designs and Patents Act 1988.

First published 2006 by
PALGRAVE MACMILLAN
Houndmills, Basingstoke, Hampshire RG21 6XS and
175 Fifth Avenue, New York, N.Y. 10010
Companies and representatives throughout the world

PALGRAVE MACMILLAN is the global academic imprint of the Palgrave Macmillan division of St. Martin's Press, LLC and of Palgrave Macmillan Ltd. Macmillan® is a registered trademark in the United States, United Kingdom and other countries. Palgrave is a registered trademark in the European Union and other countries.

ISBN 13: 978–1–4039–9150–8
ISBN 10: 1–4039–9150–2

This book is printed on paper suitable for recycling and made from fully managed and sustained forest sources.

A catalogue record for this book is available from the British Library.

Library of Congress Cataloging-in-Publication Data
Sauber, Tim, 1978–
  Structured creativity: formulating an innovation strategy / by Tim Sauber and Hugo Tschirky.
    p. cm.
  Originally presented as the first author's doctoral thesis
  Includes bibliographical references and index.
  ISBN 1–4039–9150–2
  1. Creative ability in business—Management. 2. Technological innovations—Management. 3. Strategic planning. I. Tschirky, Hugo, 1938– II. Title.
HD53.S28 2005
658.4′0714–dc22   2005047742

10 9 8 7 6 5 4 3 2 1
15 14 13 12 11 10 09 08 07 06

Printed and bound in Great Britain by
Antony Rowe Ltd, Chippenham and Eastbourne

Contents

List of Figures	vii
Acknowledgements	xi
Introduction: Management Summary	xiii

1 Introduction — 1
  Research focus and relevance — 1
  Research design and methodology — 7

2 State of the Art in Theory — 12
  Innovation strategy formulation literature — 12
  Complementary literature — 47
  Conclusion: Innovation strategy formulation (theory) — 55

3 State of the Art in Practice — 57
  Interviews — 57
  Conclusion: Innovation strategy formulation (practice) — 60

4 The Dual Gap — 61
  Dual gap in innovation strategy formulation — 61
  Working hypotheses — 62

5 Concept — 66
  Innovation architecture — 67
  Innovation strategy formulation process — 88

6 Action Research — 130
  Selection and procedure of the action research cases — 130
  Case 1: Toll Revenue — 132
  Case 2: TecChem — 138
  Case 3: HighTec — 143
  Case 4: Info Exchange — 154
  Case 5: Optic Dye — 161
  Case 6: Built-up — 167
  Case 7: RubTec — 172
  Case 8: MicroSys — 181
  Case 9: StockTec — 187
  Cross-case analysis and conclusion — 193

7 Conclusion and Working Hypothesis Rethought — 199

8 Towards a New Set of Management Principles	**203**
Understanding the innovation system	204
Formulating an innovation strategy	206
Implementing an innovation strategy formulation process	209
Notes	212
Bibliography	217
Index	227

List of Figures

1.1	Importance of innovation rated in a survey with CEOs	2
1.2	Importance of innovation strategy related in a survey with CEOs	4
1.3	Structure of this book	9
2.1	Structure of Chapter 2	13
2.2	The three basic tasks of management: design, direct and develop	15
2.3	Strategic management of a complex, systemic interacting and evolutionary system	19
2.4	Evaluation of strategy formulation schools	23
2.5	Evolution of the term 'innovation' over time	25
2.6	Comparative overview of innovation types with differing degrees of novelty	29
2.7	Integrated innovation: understanding innovation categories	33
2.8	Value creation is the addition of defining value and providing value	36
2.9	Evaluation of innovation strategy definitions	40
2.10	Integration of technology issues into strategic business planning	45
2.11	Evaluation of innovation strategy formulation concepts	46
2.12	Architecting = creating an architecture	50
2.13	The context-specific use of the architecture concept	52
2.14	The types of architecture	54
4.1	Working hypothesis 1: understanding the complex systemic interactive and evolutionary system	63
4.2	Working hypothesis 2: innovation architecture as innovation strategy formulation support	63
4.3	Working hypothesis 3: implementation of innovation strategy formulation process; company specific implementation.	65
5.1	Innovation architecture	68
5.2	Three knowledge dimensions	70
5.3	Object knowledge dimension	71
5.4	Two-level progression of idea generation	75
5.5	Functional tree	76

5.6	The definition of the functions is related to the product properties	78
5.7	The definition of the functions relates to the technology decision-making perspective	79
5.8	The definition of the functions relates to companies' strategic intentions	80
5.9	Innovation fields	81
5.10	Methodological knowledge dimension	83
5.11	Example of an innovation architecture including object and methodological knowledge	84
5.12	Major results of the innovation architecture	86
5.13	Design guidelines for developing an innovation architecture	86
5.14	Value defining processes in a company	89
5.15	Description of the virtual company Pixel AG	91
5.16	Visualization of the opportunity landscape	93
5.17	First draft of an innovation architecture at Pixel AG	94
5.18	Actual innovation architecture at Pixel AG	96
5.19	Structured and complete innovation architecture with all innovation opportunities at Pixel AG	99
5.20	The functional handshake analysis at Pixel AG	102
5.21	Key success factor analysis at Pixel AG	103
5.22	Core competence analysis at Pixel AG	105
5.23	Scenario technique at Pixel AG	106
5.24	Market portfolio analysis at Pixel AG	108
5.25	Knowledge gap analysis at Pixel AG	110
5.26	Innovation roadmap at Pixel AG	112
5.27	Resource allocation at Pixel AG	113
5.28	Establishing R&D projects' net present values	114
5.29	Dynamic technology portfolio at Pixel AG	115
5.30	Reasons for make or buy/keep or sell	117
5.31	Definitive innovation architecture at Pixel AG	120
5.32	Innovation strategy content including a simplified example of Pixel AG	122
5.33	The innovation strategy of Pixel AG	124
5.34	Deriving the innovation processes, based on the innovation architecture at Pixel AG	125
6.1	The nine action research cases at a glance	131
6.2	Project objectives of action research case at Toll Revenue	134
6.3	Innovation architecture at Toll Revenue	136

6.4	Project objectives of action research case at TecChem	140
6.5	The innovation architecture at TecChem (simplified)	141
6.6	Project objectives of action research case at HighTec	145
6.7	Innovation architecture at HighTec (Division One): segmentation by function	146
6.8	Innovation architecture at HighTec (Division Two): segmentation by products	148
6.9	Innovation roadmap at HighTec (Division One)	152
6.10	Innovation roadmap at HighTec (Division Two)	153
6.11	Project objectives of action research case at Info Exchange	156
6.12	Innovation architecture at Info Exchange	158
6.13	Project objectives of action research case at Optic Dye	162
6.14	Innovation architecture at Optic Dye (simplified overview)	164
6.15	Innovation roadmap at Optic Dye	165
6.16	Project objectives of action research case at Built-up	169
6.17	Innovation architecture at Built-up	170
6.18	Project objectives of action research case at RubTec	173
6.19	General innovation architecture at RubTec	175
6.20	Identification of new business fields based on innovation fields at RubTec	177
6.21	Innovation roadmap at RubTec	178
6.22	Technology portfolio at RubTec	179
6.23	Project objectives of action research case at MicroSys	183
6.24	Innovation architecture at MicroSys	184
6.25	Project objectives of case at StockTec	188
6.26	Deriving the innovation processes based on the innovation architecture at StockTec	190
6.27	Innovation processes at StockTec	192
6.28	Cross-case overview of project objectives and management satisfaction	194
7.1	Evaluation of the innovation strategy formulation process compared to the existing concepts	200

Acknowledgements

This book is the result of research activities at the ETH Center for Enterprise Science at the Swiss Federal Institute of Technology, Zurich. It was a great pleasure to add a book to what is known as management science. Although my name is listed as author, numerous people in academia and practice contributed to this achievement. Day to day work with these people turned out to be an enormous pleasure. They are all gratefully acknowledged.

I am greatly indebted to my thesis adviser and co-writer, Professor Dr Hugo Tschirky from ETH Zurich. His vast experience in management science and practice had a strong and positive influence on this volume. Moreover, his lively interest in and his generous support of this book allowed the research activity to be both challenging and an eventful process.

Many thanks go to Professor Dr Markus Meier from ETH, Zurich, for co-advising my book. Collaboration was a great pleasure, and his valuable input certainly improved the quality of this work.

I am also grateful to my friends and research colleagues at the ETH Center for Enterprise Science for their support during my research activity: Dr Valerie Bannert, Dr Beat Birkenmeier, Dr Harald Brodbeck, Dr Philip Bucher, Silvia Enzler, Jean-Philippe Escher, Dr Hans-Helmuth Jung, Valerie Keller, Dr Stefan Koruna, Dr Martin Luggen, Christine Müeller, Dr Pascal Savioz, and Dr Gaston Trauffler. The fact that this book appears in English is a testament to my proof-reader Hilda Fritze-Vomvoris. She is gratefully acknowledged. Many thanks also go to numerous students in our department and junior assistants in our team.

Likewise, many thanks go to Dr Thierry Lalive d'Epinay and Dr Bruno Glaus, managerial partners of the company HPO AG. From their comprehensive and well founded experience, as well as from their unconventional conceptual methods, I profited enormously. They offered me optimal part-time work on projects that allowed me to gain insight into the practice, which was highly relative to the subject of this book. At the same time, I was provided optimal free space where I could concentrate on this work. This opportunity is gratefully acknowledged. Furthermore, I wish to thank my HPO colleagues, Dr Andreas Mitterdorfer, Dr Denise Schaad, Claude Stadler, Stefan Zirhan, Martin

Stecher, Olivier Darbre, Dr Oliver Kohler and Luzia Aldighieri for their support and discussions during my book.

Special thanks go to Dr Andreas Suter for supporting me during the process of elaborating the concept of my solution. His voluntarily collaboration allowed me to integrate valuable inputs, which certainly improved the quality of this work.

The laboratory of this research was industry. The existence of this book would not have been possible without collaboration with numerous interview and action research partners. The real credit for this work goes to them.

My most sincere thanks go to my family and to Gaby Grüen, without whose great support and encouragement I would never have been able to complete my doctoral work.

Zurich TIM SAUBER

The authors and publishers would like to thank the following for kind permission to reproduce copyright material: Schaffer-Poeschel (Figure 2.7) and VDI (Figure 5.4). Every effort has been made to contact the copyright-holders, but if any have been inadvertently omitted the publishers will be pleased to make the necessary arrangement at the earliest opportunity.

Introduction: Management Summary

Innovation is vital for competitive advantage and the long-term success of a company. This means that a company has to have the ability to master future fundamental changes. However, at the same time, companies often do not have the capability to excel in innovation and have difficulty in dealing with these changes effectively and efficiently. Due to this alarming deficiency, companies are taking a broader interest in the domain of innovation management. In particular, effective and efficient strategic decision making is of major interest, the aim being to increase the accuracy of innovation objectives and the paths to reach them. In short, a concept for formulating an innovation strategy is required.

Although management literature points out the importance of innovation strategy, there is no practice-oriented, detailed and structured concept of formulating an innovation strategy, which represents a gap in the literature.

The aim of this book is to make a major contribution towards closing this gap. Therefore, the pertinent question is: How could a concept for the formulation of a structured innovation strategy be designed and implemented for innovation driven enterprises?

Serving as guidelines, working hypotheses are formulated on the basis of a literature survey. This book adopts a two-stage research procedure. First, based on interviews and existing theory, a process for formulating innovation strategy is developed. This process is based on a novel tool – innovation architecture. Innovation architecture can be seen as a blueprint for the innovation system of any company in terms of complexity, systemic interaction and evolution, which should be the basis for good management and decision making. Furthermore, innovation architecture is useful a tool for encouraging creativity on a strategic level, both in technology and business innovations, and organizational innovations. Integrating innovation architecture into the designed process is the basis for supporting practitioner oriented management in the innovation system. Second, the designed innovation strategy formulation process and the novel innovation architecture are implemented and validated in nine very different innovation driven enterprises. The innovation architecture fosters the structuring of creativity in a company.

Based on the generated and implemented elements of the innovation strategy formulation process, the working hypotheses are discussed, and an extensive set of management principles is presented in order to achieve a contribution to closing the gaps in management theory and innovation driven enterprises.

1
Introduction

Research focus and relevance

Innovation, understood as the first successful commercialization of something new, is considered beyond controversy as a key for competitive advantage and the long term success of a company (1998: 246). This significance of innovation was recently analyzed in a quantitative survey in practice by Haapaniemi (2002: 1): 'CEOs feel that innovation is critical to achieving competitive advantage ... More than 50 per cent of respondents said innovation is one of the five most important factors in building competitive advantage, and more than 10 per cent said it is the single most important factor. Executives in communication and high-tech industries, and those at companies with international operations, considered innovation especially important' (see Figure 1.1).

However, in the same study only one in ten of the CEOs strongly agreed that their organization excelled at innovation (Haapaniemi, 2002: 1). This alarming statement indicates an inability to excel in innovation, in practice.

This deficiency in innovation is aggravated by the trend towards the increasing importance of innovation, which has been confirmed by several authors.[1] They relate the ability to master future fundamental changes with the ability to develop innovations. Representative examples of such future fundamental changes include:

- **Technological changes**, which are increasing the complexity of technologies (Iansiti, 1998: 2) and are increasing the dynamic of the technological environment (Tushman and Anderson, 1997: 43)

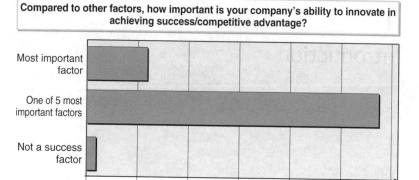

Source: Haapaniemi (2002: 1)

Figure 1.1   Importance of innovation rated in a survey with CEOs

- **Cultural and social changes**, which are creating a greater awareness of sustainable development in terms of environmental, social and governmental issues (Schofield and Feltmate, 2003)
- **Economic and industrial changes**, which are identified by the globalization and saturation of existing markets (Hamel, 2002; Nadler, 1994). At the same time product life cycles are declining (Booz et al., 1991: 26).

These changes are multifaceted and appear in an increasing dynamic (Backhaus and Zoeten, 1992: 2025). In this context Braun (1994: 122) noted succinctly: 'at present, more happens in five years than has happened in the past fifty years'.

The effect of these changes is that companies – settled in the past in continuously growing markets with stable conditions – have experienced falling sales and returns (Sommerlatte, 1987a: 5). They often reacted with organizational and cost cutting measures which, at least temporarily, improved their profitability (Call and Vöelker, 1999: 58). This reaction to cost cutting focused the companies more on saving the present than saving the future (Sommerlatte, 1987a: 5), and therefore an innovation gap arose. To solve the problem of this innovation gap, companies have to become more effective[2] at innovation (Zahn, 1995: 17), which represents the first challenge to master. In order to become more effective, it is important to consider near incremental innovations, especially radical innovations, because their return on average is much higher, while the flop rate is the same, as Berth (2003: 18) found out in a long term survey. Effectiveness of innovation is not the sole consideration; the costs and necessary time for innovation development are increasing. Thus, efficiency[3] in developing innovations is the second challenge in mastering the changes (Zahn, 1995: 17).

To improve the ability to develop innovation in an effective and efficient manner, individuals and organizations in a company must first be aware of changes in order to realize renewals in time to anticipate the situation. This means, in other words, that companies have to ensure a high degree of innovativeness. In this context, Zahn and Weidler (1995: 359) claim that a company has to be aware in order to develop innovations, in terms of successful renewals, in three different dimensions: business, technological and organizational related innovations. In so doing, the resulting innovations will have a broader range, additional protection against imitation, and thus companies gain sustainable competitive advantage by differentiating themselves from others. Such innovations are called 'integrated innovations'.

To develop 'integrated innovations', an innovation management approach is required that considers the company renewals from a holistic point of view (Zahn and Weidler, 1995: 358f). According to Hauschildt (1997: 25), individual innovation processes should be developed by anticipative[4] innovation management. This means that innovation management is part of business management (Maurer, 2002: 55). Therefore, innovation management, in the context of business management, is characterized by two questions (Hauschildt, 1997: 29):

1 What are the strategic decisions to take – in the form of an innovation strategy – in the context of innovation activities?
2 What are the consequences of the innovation strategy for the organization, based on the defined innovation strategy?

In other words, innovation management is concerned with two major challenges: firstly, with the formulation of an innovation strategy (see question 1); and, secondly, with the design of an innovation organization, including the processes (see question 2) utilized to realize the innovation intentions. In this work, innovation management has been split into two parts – innovation strategy formulation and organizational design. In this way, focus can be placed on innovation strategy formulation. Designing an organization is not the focus of this work. For a more detailed discussion of organizational design in the context of innovation management, see the thesis written by Schaad

(2001) at the Institute of Enterprise Science at ETH, Zurich. Although this work focuses on innovation strategy formulation, the results are compatible with Schaad's concept of organizational design ensuring a consistent innovation management concept.

The Focus on innovation strategy is undertaken against the background of a survey (Kambil, 2002: 8), which found that nearly 40 per cent of the interviewed CEOs indicate a clear innovation strategy as one of the most important success factors in commercializing an idea (see Figure 1.2). At the same time, the survey results indicated 'that many [companies] lack a clear innovation strategy including a decision-making team to evaluate different ideas for commercial feasibility'. According to the study, this lack of an innovation strategy affects all aspects of driving innovative ideas to commercialization; in particular, it has the greatest impact on allocating the necessary resources for innovation.

The importance of innovation strategy was confirmed in interviews undertaken in the context of this work. The interviews were conducted in several innovation driven[5] companies active in different industries. In these interviews, in addition to the problems resulting from the lack of an innovation strategy, the executives described the lack of an innovation strategy as a concern, particularly because it often leads to innovation activities not being prioritized. This results in a situation where different organizational groups work independently towards the

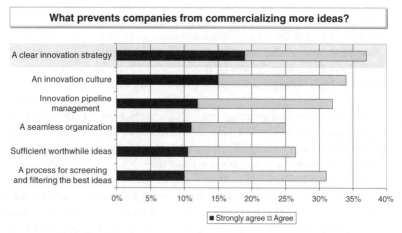

Source: Kambil (2002: 8)

Figure 1.2   Importance of innovation strategy related in a survey with CEOs

same strategic goal, and coordination problems occur. In particular, coordination problems between market driven activities and R&D driven activities were accentuated in the interviews. Although the importance of an innovation strategy was recognized by the executives, few – if any – had a solution for systematically developing such a strategy. This is a gap in management practice.

Furthermore in literature, several authors[6] point out the importance of a clear innovation strategy. From their point of view, an innovation strategy allows clear, innovation relevant, strategic goals to be set, and paths to be planned and realized in the innovation processes. Although the literature points out the importance of innovation strategy, there is no practice-oriented detailed and structured concept for the formulation of an innovation strategy. This lack is a gap in the literature.

Because no solution to the gap in practice was found in innovation strategy literature, the authors analyzed complementary literature. Based on the fact that the systematic formulation of an innovation strategy always needs a decision, and therefore understanding of the innovation system involved, the literature on system thinking has been analyzed. According to Malik (2001a: 136ff.), three aspects require serious attention to understand a system before taking a decision: complexity, systemic interaction and evolution. Most of the basic problems in a system are based, according to Malik, on disregard for these aspects. Therefore, these key aspects must be considered in order to understand – or, at least, recognize – the specific system conditions. In addition to these three aspects, it is important to develop a focusing and verbalizing model of the reality, because the reality is too multifaceted to be understood as a whole (Malik, 2002: 175). In a nutshell, to understand the innovation system in terms of innovation strategy, it is important to develop a model that considers complexity, systemic interaction and evolution.

Such a model could be described as an architecture, according to Rechtin and Meyer (1997). An architecture is a structure of a system visualizing existing and future elements and their interactions, with the aim of taking balanced decisions.

Applying the concept of architecture for the innovation system of an innovation driven company, Schaad (2001) already proposes, in the context of organizational design, the so-called innovation architecture. This innovation architecture shows the innovation-relevant connections between markets, products, technologies and scientific knowledge in a company. However, this architecture only allows a visualization of the complexity and interactions of the elements in an innovation system. The modelling of the evolution of this concept is not mentioned in this innovation architecture, which is a gap in the literature.

Nevertheless, Schaad's (2001) innovation architecture is a promising basis for developing an innovation architecture representing the complexity, systemic interaction and evolution of an innovation system. A more detailed discussion on architectures can be found in Chapter 2.

An innovation strategy is seen in management practice and literature as an important building block of innovation management. However, many companies lack an innovation strategy, and literature can not provide a structured, practice-oriented, detailed formulation concept. Nor can they provide a solution for understanding an innovation system in terms of complexity, systemic interaction and evolution as the basis for development of an innovation strategy. Therefore, the interest in the systematic formulation of an innovation strategy is an unanswered concern in management practice and literature.

The goal of this book, therefore, is to find solutions for the concerns in practice and to fill the gap in the literature:

> The research topic of this work is a concept of **structured innovation strategy formulation and its implementation**, and the research objects are **innovation driven enterprises**.

In order to be able to handle the topic, it is first of all necessary to understand the innovation system of an innovation driven enterprise. Therefore, the first research question is:

> How can a **complex, systemic interactive and evolutionary innovation system** be modelled in order to **understand** the system specific conditions of an innovation driven enterprise?

The results of this first research question should be a model of a complex, interactive and evolutionary innovation system, the so-called innovation architecture. This innovation architecture has to be integrated, in a second step, into a strategy formulation concept for implementation in practice. Therefore, the second and third research questions are:

> How could a structured innovation strategy formulation concept be **designed**, based on the innovation Architecture, for innovation driven enterprises?

How could such an innovation strategy formulation concept be **implemented**?

In fact, the design of a concept and its implementation cannot be strictly separated from one another. The emphasis is, by the nature of organizational research, on the organization of structure, processes, and methods of innovation strategy formulation in innovation driven enterprises.

Research design and methodology

Structure of the book

The emphasis of this work is dual faceted; firstly, it is on gaining insight into business reality by means of first-hand information in order to develop solutions that are of use to practitioners. Secondly, this solution is implemented in practice by application-oriented research. The output is a set of management principles rather than a simple description of concepts for innovation strategy formulation. It is, therefore, a practitioner-oriented concept of an innovation strategy formulation concept, based on a model representing the systems-specific conditions of an innovation driven enterprise. This concept should represent a holistic approach to innovation strategy formulation.

Because of the significant shortcomings in the literature, an empirical and explorative research design is initially preferred. Most current empirical and explorative studies are based on case studies. The choice of the case study research method is explained by Kubicek (1975: 61). He argues that case studies are best for the very early stages in research of an organizational problem. They require relatively little effort and bring plenty of suggestions for further research on this topic. However, Lang (1998a: 132) points out some weaknesses of case studies, which are mainly that they provide an inadequate generalization of insight.

With regard to this concern, a research procedure has been chosen for this book. It is based on Yin's (1994: 49) multiple case study approach. In the first stage, a theoretical concept of innovation strategy formulation is designed and several cases (innovation driven enterprises) are selected. In the second stage, the theoretical concept is evaluated in terms of action research cases[7] in the different enterprises. The following analysis is an interpretation of the action research case results as a two level concept,[8] which provides an individual analysis of each case and then provides a cross-case overall analysis. Following this two level evaluation, the theoretical concept is modified. On the basis of the collected insight, it is possible to derive management principles that allow insight gained to be transferred to other business cases, and therefore enables them to be multiplied. By this procedure 'every case should serve a specific purpose within the overall scope of inquiry' (Yin, 1994: 45).

This book does not, however, claim an overall validation of the learning from the action research cases. This is not possible for two reasons. On one hand, the sample of action research case studies is always restricted by the number of companies studied; on the other, the insight gained from action research case studies always reflects one company's specific reality. Thus, the aims of this research are to explore how an innovation driven enterprise can develop an innovation strategy and implement its formulation concept, and then to present management principles for guiding an innovation driven enterprise to an appropriate solution.

The inclusion of the action research cases is the result of close research cooperation between the author and the companies. These action research cases should, through exploration, build a basis for the discussion of the elements of a concept for innovation strategy formulation. Due to the fact that is was impossible to find one company that could provide the basis for the full range of decisions, the authors worked with nine different companies. Subsequently, lessons learned from empirical research flow into a set of management principles that guide innovation driven enterprises to design and implement an innovation strategy formulation concept. The structure of this book is depicted in Figure 1.3, which gives an overview of the major parts in this work.

Action research

The term 'action research' is attributed to Lewin (1946). His work seems to be fundamental to the modern understanding of action research: 'He created a new role for researchers and redefined criteria for judging the quality of an inquiry process. Lewin shifted the researcher's role from being a distant observer to involvement in concrete problem solving' (Greenwood and Levin, 1998: 19). Since the 1970s, Kubicek (1975) has observed an intensified attention to action research and he paraphrases the term with research by development – in organizational research.[9] He designates action research as an approach in which practitioners and scientists jointly design and implement new organizational concepts. Moreover, the scientists involved try in turn to systematize and generalize their experiences

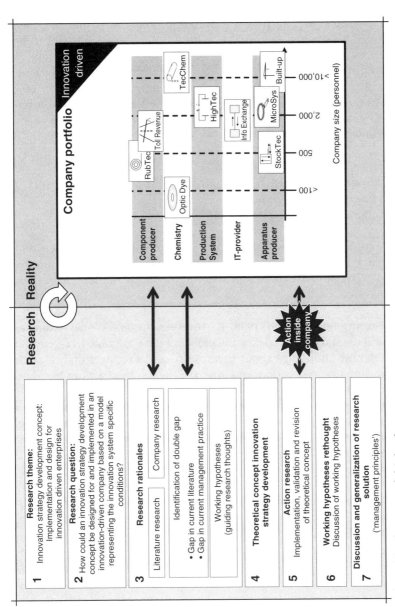

Figure 1.3 Structure of this book

(Kubicek, 1975: 70). Thus, action research is action-oriented. This means that the researcher has active influence over the research object for making changing from inside (Popp, 2001: 401) and can pass through a learning process that allows the examination field to be changed during the analysis or permits more information sources (Wollmann and Gerd-Michael, 1977: 445). In contrast, this is not possible through passive approaches such as pure case studies. In case studies, the researcher is limited to formulating questions and interpreting the empirical results. Three terms central to action research are (Greenwood and Levin, 1998: 6):

- research (knowledge generation)
- participation (a participatory process in which everyone involved takes some responsibility)
- action (jointly elaborated options of action).

According to Greenwood and Levin, one can only speak about action research in its proper sense if all of these aspects are considered in the study. To clarify the content of action research, two current definitions are given:

- Greenwood and Levin (1998: 4): 'Together, the professional researcher and the stakeholders define the problems to be examined, co-generate relevant knowledge about them, learn and execute social research techniques, take action, and interpret the results of actions based on what they have learned.'
- Cunningham (1993: 4): 'Action Research is a term for describing a spectrum of activities that focus on research, planning, theorizing, learning, and development. It describes a continuous process of research and learning in the researcher's long-term relationship with a problem.'

Facing the often cited contrast of qualitative and quantitative research methods, action research adopts a somewhat neutral position. In principle, action research allows any kind of method of social science. 'Surveys, statistical analysis, interviews, focus groups, ethnographies, and life histories are all acceptable, if the reason for deploying them has been agreed on by the action research collaborators and if they are used in a way that does not oppress the participants' (Greenwood and Levin, 1998: 7). Thus, action research seems to be very promising for exploratory studies in organizations. Both, the

Introduction 11

research community and the organizations benefit from the experience gained during common design and implementation of new concepts. The great potential of this kind of research lies in the two supplementary dimensions of action and participation. Therefore, the action research projects in this book are based on a cooperative problem solving and learning approach. It is a tightrope walk of interests for scientists and practitioners. This is particularly challenging for the management of the action research projects conducted. In practice, nevertheless, action research is normally restricted to a few companies because of the limited degree of readiness of organizations to cooperate (Kubicek, 1975: 71). Fortunately, nine companies were prepared to support this action research.

Empirical research methodology and raw data

Non-standardized interviews have an important place in this book. This qualitative empirical approach is justified by two arguments. Firstly, written questionnaires do not make sense because of the heterogeneous use of different terms in the field of innovation strategy formulation. Secondly, activities in this field of interest seem to be very informal, and therefore cannot be mapped by standardized methods. In general, written questionnaires often fail because of the complexity of the topic, and consequently the need for clarification is significant. The major disadvantages of qualitative, empirical research designs for organizational research are that the sample's representative nature is always limited. Despite this problem, the main advantage lies in the opportunity to identify the neglected phenomena, coherence of causes and effects, processes and so on (Bortz and Döering, 1995), and, therefore, to structure a very complex subject on the one hand, and on the other, to bring new aspects to the surface. A detailed description of the methodologies (interviews, workshops, document analysis and so on) adopted for each stage, and indications for the sources of raw data will always be presented at the beginning of each section. The author of this book is committed to honest and transparent research. Therefore, the origin of the raw data for each case it is clearly stated. In addition, the author does not wish to adorn himself with borrowed plumes. Raw data is part of projects, diploma and post-diploma theses at the ETH Center for Enterprise Science; that is, Finckh (2003), Flüehmann (2003), Lincke (2004). They are all gratefully acknowledged for their collaboration.

2
State of the Art in Theory

Chapter 1 showed the importance of research into the concept of a structured innovation strategy formulation based on a model representing the innovation system of a complex, systemic interacting and evolutionary innovation driven company. A brief outline of the literature was presented. In contrast, this chapter introduces in detail the literature of the research topic, together with complementary literature, and concludes with the gap in innovation strategy formulation.

Innovation strategy formulation literature

The aim in this chapter is to introduce the research subject in detail by showing representative concepts of the literature and the essential criteria to consider in each domain. Four domains are presented: (1) strategic management, (2) strategy and strategy formulation, (3) innovation and innovation management, and (4) innovation strategy and innovation strategy formulation. These domains are presented against the background of the innovation strategy formulation, which is firmly based on the understanding of these subjects. Therefore, the author has chosen a constructive bottom-up approach to introduce innovation strategy formulation. This bottom-up approach is the underlying structure of this chapter (Figure 2.1).

After introducing strategic management, strategy and strategy formulation, the essential criteria of these domains are identified and strategy formulation concepts are evaluated based on these criteria. This first preliminary evaluation is undertaken on the basis that the research in literature on innovation strategy formulation concepts can be focused on concepts that are consistent with the understanding of general strategy formulation in this work. After the preliminary evaluation of

State of the Art in Theory 13

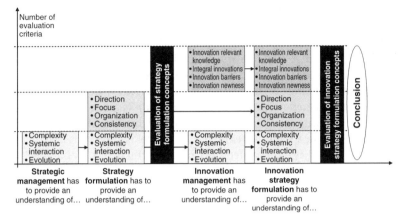

Figure 2.1 Structure of Chapter 2

strategy formulation concepts, the terms 'innovation' and 'innovation management' are introduced, which leads to innovation strategy and innovation strategy formulation. With the additional identified innovation-specific criteria, the innovation strategy formulation concepts in literature are evaluated and a conclusion is drawn with the aims of identifying interesting concepts in the context of innovation strategy formulation and defining the subjects to be analyzed in the complementary literature.

Strategic management

The issue of strategic management is probably the most philosophical topic in management literature. As will be shown later, there are diverse schools and a series of approaches for dealing with the question of strategy. From an initial point of view, strategic management is understood as planning how to run and change the business in order to achieve business missions and goals (Wright *et al.*, 1992: 3).[1] Terms like 'thinking', 'acting', and 'decision-making' are central tasks to this purpose (Gälweiler, 1990: 65).

Strategic management tasks

All the above mentioned 'classical' tasks of management can be summarized in three basic categories: design, direct and develop. These tasks are based on three vertical levels: normative, strategic and operational (vertical integration) as well as three horizontal levels: structures, behaviours and objectives (horizontal integration). The goal is to design, direct and develop the company potentials to influence the company as well as its environmental development process.

Design aims to create a working system. For this purpose, the required resources and institutional domains will be defined and merged into a consistent organization. Design as a management function means the draft of an institutional model, for which the designation of the targeted properties of the institution is essential. Therefore, such a design can be called a 'design model', which must be clearly differentiated from a scientific explanation model (which tries to explain an existing reality) and a decision model (which displays a specific problem situation in an existing system). In contrast, design models are analogous to construction drafts, the aim being to create a reality not yet in existence. The development of this design model is therefore an eminent constructive procedure (Ulrich and Probst, 1988: 260).

Direct means that the company objectives are aligned in real time to the created objectives of the design model. This requires a constant debate with the conditions of the environment as well as with the situation of the company itself. The outcome of this is a constant need to evaluate changes to planned projects, which demand a further decision and its implementation. Direct is therefore a function, which is essential in a system so that this system can reach its objectives under changing conditions by means of concrete activities (Ulrich and Probst, 1988: 261).

Development is an executive task, which is the cognizant constant change to the system and its direction. What is essential to this task are the social, technological and industrial changes that result in changed conditions and assumptions for the design and direction of the company's systems. Thus, to the fore is the further development of the company in terms of constant improvement or qualitative learning. In the short term, it is essential that the company learns to function better with given objectives for the subsequent elimination of deficiencies and avoidance of mistakes. In the long term, it is vital to encourage the innovativeness[2] of the company (Ulrich and Probst, 1988: 263).

In Figure 2.2 the three basic tasks of management are summarized. It is apparent that design and direction are primarily concerned with thinking in continuities, whereas the development of the company requires thinking in discontinuities. The tasks of design and direction are engaged with the 'from today to tomorrow', whereas the task of development is concerned with the 'from today to after tomorrow'.

To manage a company in terms of designing, directing and developing it, it is essential to understand the company's steering values (cf. Gälweiler, 1990: 35): firstly, which are the real strategic orientation values; secondly, what constraints exist between them, thirdly, which sources provide information; and fourthly, what time frame they forsee (Malik, 2001b: 7). Gäelweiler (1990: 34) presents a concept for answering these questions by analyzing the company from a steering and regulation point of view.

Summarizing Gäelweiler's (1990: 34) concept, it is essential, on an operational level, to understand the steering values of liquidity and success and, on a strategic level, to understand that the steering values of the existing successful position and the potential for new success are of primary interest. In this context, it must be mentioned that the strategic steering values have to be taken into account in a pre-steering phase (Malik, 2001a: 143). This is because the operative implementations of specific decisions are delayed. It is therefore essential to pre-steer a system, such as a company, to be able to implement decisions suffiently early. With this understanding of pre-steering for the company, the strategic steering value 'the existing success position' is the basis for designing and directing a company in the context of the concept shown in Figure 2.2, whereas the strategic steering value 'new success potential' is crucial for developing a company. On the strategic

'From today to tomorrow' – primarily: thinking in continuities

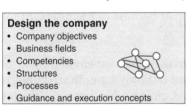

Design the company
- Company objectives
- Business fields
- Competencies
- Structures
- Processes
- Guidance and execution concepts

Direct the company
- Guidance of company and human resources
- Execution of individual performance and collective performance

'From today to after tomorrow' – primarily: thinking in discontinuities

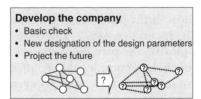

Develop the company
- Basic check
- New designation of the design parameters
- Project the future

Figure 2.2 The three basic tasks of management: design, direct and develop

level, this leads to the main task of strategic management, which is differentiated because of the time horizon:

- Firstly, the actual system of a company and its changes, which are not structural, have to be designed and directed by using the existing success position as a strategic steering value.
- Secondly, the evolution of the actual system of a company which implies structural changes has to be developed using new success positions as strategic steering values.

The awareness of these two tasks of strategic management allows a sensible look at the essential strategic steering values mentioned above, and allows a pre-steering in terms of the strategic management in a company. But these strategic steering values still have a weakness according to Malik (2001a: 135): steering values of a company, or for every system, can systematically misguide. This is because the perceptions of the significant basic data for the steering values, originating in the specific history of the company, are not changed over time. Therefore, it is important to understand systematically – and not traditionally – the company's system for adapting strategic steering values data when pursuing strategic management. This is the subject of the next section.

Systemic understanding in strategic management

According to Malik (2001a: 135f.) a system, such as a company, is often misunderstood in strategic management because of the disregard for three major factors:

- **The complexity** of companies and their relevant environment. In this context complexity stands not only for 'complicated', but it also marks the capability of a system to adopt many different states. Complexity can be quantified and measured with the support of the term variability. 'Variability is the number of different states of a system or the amount of a set of different elements' (Malik, 1992: 186). This understanding of complexity is in alignment with authors in the domain of cybernetic research, such as Hayek (1972; 1973) and Beer (1972; 1979). To get a more detailed feeling for complexity, the playing of chess is an adequate example: The game of chess is considered as complex because an enormous number of different moves and an even higher number of possible configurations on the chess board is possible. This number of possibilities can be handled neither by a human brain nor by a computer. Therefore, this game has a strategic character.

- **The systemic interaction**[3] between the determining influential elements understood as variables, thereby systemic meaning 'pertaining to a system or systems' (Davis, 1980: 705). Therefore, a systemic interaction means that the relations between elements pertain to a system and that they therefore have an influence on the system. The interactions are normally not only linear and monocausal, but also positive or negative back-coupled. If, in these interactions, a threshold of values is exceeded, often the character of an element will change by generating an overturn of an effect or a step function. These changes make it difficult to evaluate and interfere with a system. Therefore, it is essential to analyze the systemic interactions of elements to be able to predict, although still with a definite uncertainty, the side effects.
- **The evolution of a system.** Based on evolutionary theories,[4] evolution is understood as constant, often unpredictable changes that, in the long term, have the effect of provoking structural changes in a company (Malik, 2001a: 137). The evolution is based on the multilateral action and reaction of elements in a complex, systemic interacting system, which consistently provokes new situations. These situations are, in general, the result of the interactions between two organisms that have different objectives and are therefore acting reciprocally. During the reciprocal action, a process of adaptation is activated that is cognizable in an increment in knowledge of the organisms (Bartley, 1987: 23). The cause of this adaptation is the dynamism of a system. Only if it is possible to identify the underlying objectives and processes of each element is it possible to orient them in one direction. It is essential to identify the process of adaptation, which is seen by the development of the knowledge in the environment. It is then possible in those circumstances to realize certain fixed basic structures in the future.

According to Malik (2001a: 138), these three criteria are the starting point for understanding a system. How the understanding of these fundamental criteria are dependent on the strategic steering values and strategic management tasks previously mentioned is further discussed in the next section.

Conclusion

As previously mentioned (pp. 14ff.), an actual system of a company can be designed and directed by using the existing success position as a strategic steering value. To understand the strategic steering value in a pre-steering phase, it is relevant first to understand the complexity and the systemic interaction of a company understood as a system. In contrast, the understanding of evolution supports the derivation of new success positions and the develpoment of a future system. This relation between understanding a system, defining the steering values and accomplishing strategic management tasks is shown in Figure 2.3.

To bring it all together in a pre-steering phase, it is essential for strategic management first to understand the system in terms of complexity, systemic interaction and evolution. Therefore, this understanding of the system is, according to Malik (2001a), a basic criteria for the creation of a strategic management concept: **strategic management has to design, direct and develop a company based on a clear understanding of company systems' complexity, systemic interaction and evolution.**

Based on this understanding of strategic management, the terms 'strategy' and 'strategy formulation' will be discussed in the next section.

Strategy and strategy formulation

Strategy

A system, such as an electrical circuit, can be pre-steered in its voltage, because in the case of a reaction to an environmental effect, it is too late to steer the circuit in real time. Therefore, an attempt is made to anticipate future influences of the environment by sending clear signals to the electrical circuit in terms of pre-steering. These signals contain a clear objective to decrease or increase the tension, and a path, by means of a current transformer.

Analogous to the pre-steering of an electrical circuit, the system of a company can be pre-steered. A company tries to anticipate future influences from its environment in order to decide on a strategic management level, to implement a decision by submitting a signal to the operational level of the company. This signal contains a clear objective and a comprehensive path, without defining in detail what to do at every step. More precisely, this pre-steering signal is, in general, and in company terms, a strategy.

However, in this description of strategy the term varies in its definition from author to author, and it has become a key word in management practice and literature: 'For a term so central, one even incorporated in the name of the field, it might be expected that a common definition would exist for the word strategy. Yet the concept remains different for different users' (Schendel and Cool, 1988: 23).

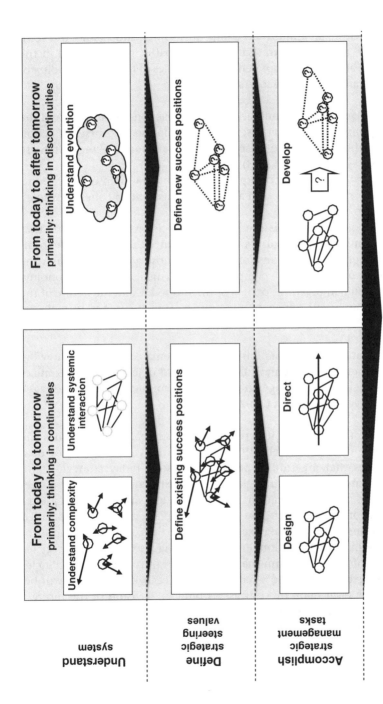

Figure 2.3 Strategic management of a complex, systemic interacting and evolutionary system

Nevertheless, an important group of authors understand the term 'strategy' as defining strategic goals and paths, as mentioned in the example above. Some representative examples of this understanding of strategy are listed in the following:

- 'Strategy is the determination of the basic long-term goals and the objectives of an enterprise, and the adoption of courses of action and the allocation of resources necessary for carrying out these goals': Chandler (1962: 13)
- 'A strategy is a plan or a pattern that integrates an organization's major goals, policies, and action sequences into a cohesive whole': Quinn (1980: 7)
- 'Corporate strategy is the pattern of decisions in a company that determines and reveals its objectives, purposes, produces the principle policies and plans for achieving those goals, and defines the range of business the company is to pursue, the kind of economic and human organization it is or intends to be, and the nature of the economic and noneconomic contribution it intends to make its shareholders, employees, customers, and communities': Andrews (1987: 13).
- 'A strategy of a corporation is a comprehensive plan stating how the corporation will achieve its mission and objectives. It maximizes competitive advantage and minimizes competitive disadvantage': Hunger and Wheelen (2002: 7)

Beside this distinction between path and goal, which defines the strategic direction, Abell (1999) suggests the consideration of dual strategies that are run in parallel: 'today-for-today strategies' and 'today-for-tomorrow strategies'. 'This distinction between a present and future orientation is not the usual short-term, long-term distinction – in which the short-term plan is simply a detailed operations and budgeting exercise made in the context of a hoped-for long-term market position. Present planning also requires strategy – a vision of how the firm has to operate (given its competencies and target markets) and what the role of each key function will be. The long-term plan, by contrast, is built on a vision of the future – even more important, on a strategy for getting there' (Abell, 1999: 74). This dual term understanding of strategy is in alignment with the understanding of strategic management to consider today, tomorrow and after tomorrow based on the understanding of complexity, systemic interaction and evolution of a company.

State of the Art in Theory   21

Therefore, to summarize, a **strategy sets strategic goals and strategic paths considering the today, the tomorrow and the after tomorrow for setting the company direction, focusing efforts, allowing the organization to be defined and providing consistency in a balanced manner.** This represents the understanding of strategy in this book.

Strategy formulation

Up to this point, the understanding of strategy in this work is clear. Nevertheless, there are varying ways of formulating strategies. Mintzberg and Lampel (1999) have analyzed different strategy formulation concepts and have clustered them into ten different basic schools of strategy formulation. These schools emphasize that, during the last decades, the understanding of strategic management underwent a great change, certainly based on the insight that it is less possible to plan the future than expected. The ten strategy formulation schools are: the Design School, Planning School, Positioning School, Entrepreneurial School, Cognitive School, Learning School, Power School, Cultural School, Environmental School and Configuration School.

Across these ten schools of strategy formulation, various approaches become a hybrid by linking some or all of the elements of the different schools (Mintzberg and Lampel, 1999: 26). Examples are: stakeholder analysis (linking the planning and positioning schools), chaos theory (being a hybrid of the learning and environmental schools), dynamic capabilities (being a hybrid of the learning and design schools) and resource-based theory (being a hybrid of the learning and cultural schools).

Based on these different strategy schools, the emerging question is whether they represent different strategy processes or complementary parts of the same process. Mintzberg and Lampel (1999: 27) come to the conclusion that most schools partly represent aspects of what can be referred to as strategy formulation, which is in sum 'judgmental designing, intuitive visioning and emergent learning; it is about transformation as well as perpetuation; it must involve individual cognition and social interaction, cooperative as well as conflictual; it has to include analyzing before and programming after as well as negotiating during; and all this must be in response to what may be a demanding environment' (Mintzberg and Lampel, 1999: 27).

Although Mintzberg and Lampel point out that all aspects in strategy formulation should be considered, there are approaches that are more in alignment with the understanding of strategic management and strategy of this work than others. Therefore, in the next section the ten schools will be evaluated in order to identify the schools that tend to be a basis for this book.

Evaluation and conclusion

The evaluation criteria for evaluation of the ten strategy formulation schools are based, on one hand, on the basic strategic management criteria for understanding a company in terms of complexity, systemic interaction and evolution (p. 10) and, on the other, the understanding of the term 'strategy' setting direction, focusing efforts, allowing the organization to be defined and providing consistency in a balanced manner. The evaluation of the ten schools with these seven criteria is shown in Figure 2.4 and a discussion follows.

Fulfilment of specific criteria for strategic management: In Figure 2.4 it can be seen that there is no approach that explicitly fulfils the three criteria of strategic management. Only three schools implicitly support the strategic management criteria. The design school and planning school, which are very similar, have a procedure with detailed process steps and checklists for understanding the company. This could permit a good understanding of the company as a whole. Nevertheless, this procedure is not focused on an understanding of complexity, systemic interaction and evolution. Therefore, these two schools can only implicitly provide an understanding of the three strategic management criteria. The cognitive school does try to understand the company with models but does not explicitly try to understand the three strategic management specific criteria.

Fulfilment of specific criteria for strategy: These four criteria are not fulfilled by any one school completely. There are five schools that very nearly fulfil all the strategy specific criteria: the design school, planning school, positioning school, cognitive school and configuration school. These five schools are characterized by the fact that they have a similar understanding of a strategy. This signifies that the direction is clear, the focused efforts are defined and consistency is ensured. Despite this clear statement of a strategy, there is no strategy formulation process that gives a clear understanding of how a strategy should allow definition of an organization.

The design, planning and cognitive schools have the most common approach to how a strategy formulation concept should be designed according to the seven criteria. The design and planning school are the most interesting because of their systematic checklists, and the cognit-

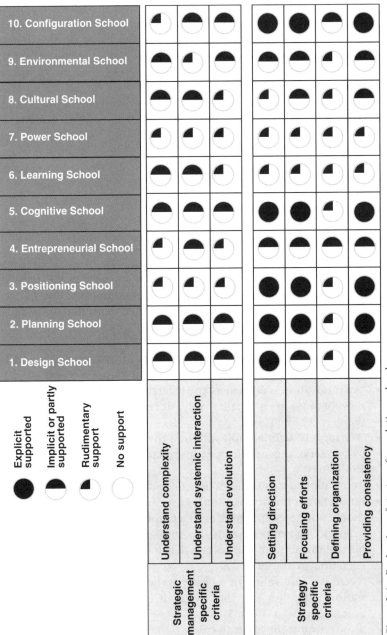

Figure 2.4 Evaluation of strategy formulation schools

ive school is a good basis with its concept for modelling a system. Therefore, in the further sections in the innovation specific context only strategic concepts are presented that are based on the design, planning or cognitive schools.

Innovation and innovation management

Before detailing the term 'innovation strategy' and describing possible innovation strategy formulation processes, it is first of all essential to describe the terms 'innovation' and 'innovation management'. This will demonstrate the essential criteria to notice in innovation management, and which criteria are essential for formulating an innovation strategy.

Innovation

'Innovation' is a modern term (Hauschildt, 1997) which, due to its multiple use in everyday speech and research, is defined and interpreted very differently. However, the term is not an invention of recent years. The application area of innovation has appeared in literature for more than half a century. The sense of the term has steadily changed during this time. In the following are selected definition approaches that document the development:

- 'If, instead of quantities of factors, we vary the form of the production function, then we have an innovation' (Schumpeter, 1939: 87)
- 'An innovation is defined as any thought, behavior, or thing that is new because it is qualitatively different from existing forms' (Barnett, 1953: 7)
- 'We suggest defining innovation as the first or early use of an idea by one set of organizations with similar goals' (Becker and Whisler, 1967: 463)
- 'Innovations are the units of technological change' (Marquis, 1969: 1)
- Innovation is the first (economic) use of an invention. The invention should not necessarily have emerged from research and development in science, but comprehends also novel objects and processes of business administration and social science in the broadest sense (translated from Witte, 1973: 17)
- 'Technical innovation in industry is the development, commercialization, adoption and improvement of product and production processes' (Pavitt, 1980: 1)
- 'Innovation is the effort to create purposeful, focused change in an enterprise's economic or social potential' (Drucker, 1985: 67)

- 'Innovation is the battle in the marketplace between innovators or attackers trying to make money by changing the order of things, and defenders protecting their existing cash flow' (Foster 1986: 20)
- 'Innovation consists of the generation of a new idea and its implementation into a new product, process, or service, leading to the dynamic growth of the national economy and the increase of employment as well as to a creation of pure profit for the innovative business enterprise' (Urabe, 1988: 3)
- 'Innovation is a new way of doing things that is commercialized. The process of innovation cannot be separated from a firm's strategic and competitive context' (Porter, 1990: 780)
- 'Innovation is defined as adoption of an internally generated purchased device, system, policy, program, process, product or service that is new to the adopting organization' (Damanpour, 1991: 556)
- 'Innovation is the purposeful implementation of new technical, economical, organizational and social problem solutions, that are oriented to achieve the company objectives in a new way' (translated from Vahs and Burmester, 1999: 1).

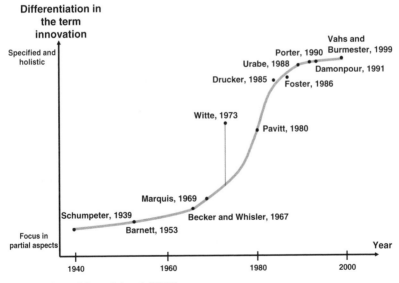

Source: adapted from Schaad (2001)

Figure 2.5   Evolution of the term 'innovation' over time

These definitions can be categorized according to their meaning. The differences are obviously due to the fact that the term 'innovation' has changed over the years. Whereas, in the past, one detail was considered, today many aspects are. Additionally, nowadays the term 'innovation' is much more success-oriented than it was in the past.

The definition by Schumpeter (1939: 87), which describes innovation as variation of factors, was still a little vague. By its time the term 'innovation' was specified in its definition, as can be seen in Figure 2.5. In this figure, the definition of Witte (1973: 17) stands out. He was, for his time, very progressive and was already very near to the present understanding of innovation.

Additionally, is it possible to see, based on the listed citations, that in the 1960s and 1970s the term 'innovation' was understood as a form of changing process.[5] In the late 1970s, the term seemed to be more refined. More and more, the preconditioned success of an innovation was part of the definition. This is reflected in the terms 'effectiveness', 'profitability', and 'customer satisfaction'. The changes of the understanding of 'innovation' are certainly influenced by the effect of new constellations in the environment that are caused through increased competition, changes in the market environment and dynamic changes in technology. A definition, appropriated today, of the term 'innovation' could be defined in agreement with actual literature according to Schaad (2001: 15):

Innovation is a first successful commercial use of something new by an enterprise

This attempt to define innovation is understood as an underlying working definition and is used in this work.

In the next sections the term 'innovation' will be analyzed from several points of view. In doing so, the authors wish to explain the term in greater detail and demonstrate its importance for companies. Also the discussion of innovation allows essential aspects to be considered in developing a strategy in the domain of innovation to be identified.

Innovativeness and barriers to innovation

To develop innovations, according to the above definition a company has to be innovative, whereas innovativeness can be understood as the ability of individuals and organizations to be aware of changes in order to realize renewals early and anticipate events. This means that a

State of the Art in Theory 27

company should be able to use their potential capabilities[6] to create new products and processes and to commercialize them (Wagner and Kreuter, 1998: 34).

For an innovation driven company, a high degree of innovativeness is therefore essential. To reach this objective, so-called innovation barriers have to be mastered. According to Bond and Houston (2003: 125) such innovation barriers can be classified in three groups. The following overview should be seen as exemplary and not as a complete overview of the literature:[7]

- **Technology and market barriers** address issues of whether a technology can be applied to meet a customer need at a profit. In this category, there are technology barriers to be mastered (such as the availability of the required technologies) and market barriers (such as developing the optimal business model to commercialize new products based on emerging technologies). These two more or less detached innovation barriers are accompanied by a more comprehensive barrier, the technology–market linkage, understood as the degree to which the technology can be matched to customer demand within a current or potential market opportunity.
- **Strategy and structure barriers** focus on the roles played by the firm's technology capabilities, strategies, and supporting organizational structures in successfully deploying a technology to meet a market need. This means that problems in innovation can occur through the lack of capabilities to develop a technology, limited resources or the absence of a link of innovations to companies and business unit strategy, and therefore the missing link to their actual portfolios. These barriers could be generated by a lack of an innovation strategy (see Figure 1.2) (Kambil, 2002); For example, the effects of a lack of strategy, where the resources are not focused and are therefore even more limited for important projects, and where important capabilities are not systematically developed with a goal-oriented focus.
- **Social and cultural barriers** account for the effects of differences in frames of reference, values and goals in matching technologies to market opportunities. These innovation barriers have their origin in the different interpretation and communication of functional specialists, whose beliefs differ (Dougherty, 1992). This innovation barrier is especially important in the context of the functional units of marketing and R&D, because they are mainly responsible for innovation. In this context, Souder (2004: 602) has identified in a

survey including 289 projects that the harmony between R&D (responsible for future technologies) and marketing (responsible for the identification of customer demand) is not working optimally. In 59.2 per cent of projects, a mild to severe disharmony between R&D and marketing was ascertained. This disharmony is an innovation barrier in terms of aligning market pulls (understood as the development of technologies due to a concrete market need) and technology pushes (understood as the development of technologies creating a new market need). The alignment of technology push and market pull is very important in developing innovations.

These various and multifaceted innovation barriers often hinder companies ability to reach a high degree of innovativeness. Furthermore, additionally to these innovation barriers, companies are facing the fact that the required degree of innovativeness to successfully persist in competition with competitors is increasing with time.[8]

To develop a high degree of innovativeness, there is no single best way, 'however there are a number of consistent themes, which provide a blueprint from which to try and configure innovative organizations' (Bessant, 2003: 35). According to Bessant (2003: 35f.), the essential factors to be aware of in order to create an innovative organization are multifaceted and interdepend on several factors. In this context, Meier *et al.* (2004: 2f.) propose a seven-fold categorization of innovation enablers: strategy, resources, processes, methods, tools, organization and culture.

It is not possible to augment innovativeness rapidly and sensible interactions in organizations have to be taken into account. In this context, Sommerlatte *et al.* (1987b: 57) points out that 'innovativeness cannot be "bred" like a truffle. It does not suffice, just when it is economically and politically opportune, wanting to turn on the innovation tap, by adapting one or another factor.'[9] The long-term horizon that is aligned with innovativeness is additionally emphasized by the following statement from Gassmann and Zedtwitz (1996: 39): 'Successful enterprises know that innovation is no lucky chance, but the result of innovativeness – the ability, continuously and systematically to develop new customer value and commercialize it.'[10]

Newness of innovation

Innovations are not always new in the same degree of novelty. As a 'real' innovation we would primarily consider the development of the steam engine, the electric motor or the telephone. But the walkman is also an innovation. Batteries, magnets and earphones already existed. The new element was the idea of portable entertainment, which allows music to be listened to anywhere. The development of the walkman is therefore a new combination of existing elements, a rearrangement. Is a cooker with a glass-ceramic surface, compared to the already existing resistance heater with the steel hob, a major or a small innovation? What about the inductive hob of the microwave oven? In this context, Abernathy and Clark (1985: 4) noted that 'some innovations disrupt, destroy and make obsolete established competence; others refine and improve'. They are different in their degree of novelty. This difference in their degree of novelty is defined by the qualitative difference of an innovation compared to the previous state (Hauschildt, 1993: 39). Innovations with a high or low degree of novelty can be distinguished by the innovation type. These so-called innovation types are defined by different authors. The principle taxonomy of the innovation types is slightly different comparing the authors' use of terminology. Some examples of this are presented in Figure 2.6.

In the area of a high degree of novelty, the definitions basically accord. In the case of the definitions for a low degree of novelty, it is important to note Mensch's (1975: 54) definition, which illustrates with the innovation type 'pretended innovation' (which is at the limit of not being a novelty) that the limit of innovations is mobile. Tushman and

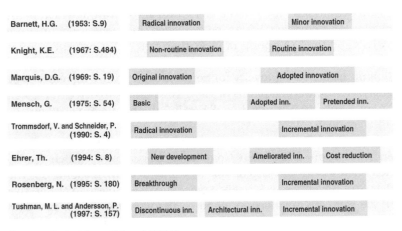

Source: adapted from Schaad (2001)

Figure 2.6 Comparative overview of innovation types with differing degrees of novelty

Anderson (1997: 157) furthermore establish the term 'architectural innovation',[11] which indicates a middle degree of novelty. The three main degrees of novelty can be summarized as follows:

- **Low degree of novelty**: ameliorations or modifications of the actual performance[12]
- **Middle degree of novelty**: new combinations based on already existing elements[13]
- **High degree of novelty**: a completely new performance.[14]

Behind these definitions of the degree of novelty are differing viewpoints. Some authors are more influenced by a technology view, such as Rosenberg (1995: 180) and Tushman and Anderson (1997: 157). The authors with a market-based viewpoint are Knight (1967: 484) and Ehrer (1994: 8). Seibert (1998: 112) has a mixed viewpoint, in which he points out that innovations with a high degree of novelty are based on a 'technology push' and should therefore be defined by a technology point of view. The more frequent lower innovations with a low degree of novelty are mainly 'market pull' and their definition should be given from a market point of view.

There is the question, for example, of how innovative the implementation of the SBB Easy-Ride-Ticket[15] is, the Swiss having already implemented E-ticketing and fast-track[16] more than three years ago. This is still, in spite of the definition of the degree of novelty, not easy to answer, because it has to be taken into account from which point of view the question has to be answered. Without a doubt, from Swiss's and the SBB's point of view, their new ticketing generation is an innovation with a high degree of novelty. The answer to this question of the degree of novelty for both innovations has therefore to be analyzed from several points of view. The point is to detect for which group a novelty is innovative. Basically, the points of view must be differentiated.

An initial differentiation is an evaluation of the degree of novelty of an innovation undertaken from inside a company, which is also described as the micro-economic[17] point of view. In doing so, the degree of novelty is exclusively defined out of a subjective perspective of the company involved. A micro-economic innovation is therefore an innovation where a change for the specific company is new. For this reason, imitations or adaptations of external developments are also considered as innovations. Therefore, it is not important how long the considered innovation has existed in other places.

A second differentiation of the degree of novelty is the macro-economic view (Kaplaner, 1986: 15). The degree of novelty is evaluated by comparing the innovation to the total offers on the market. The degree of novelty is, in this case, the highest when it is a worldwide successful renewal. A more restricted macro-economic view is proposed by Hauschildt (1993: 15); the industry-economic view. This view restricts the evaluation of the degree of novelty to the relevant environment. Therefore, a high degree of innovation can be reached if, in the same industry, there is no comparable renewal. The Swiss and the SBB, to come back to the examples of the new ticket machines, are both acting in the domain of transportation, but their environment is totally different, and therefore, their industry-economics are different. Thus, from an industry-economic point of view, both developments are innovations with a high degree of novelty.

Booz, Allen and Hamilton (1982: 9) showed through a study that, from the micro-economic point of view, 30 per cent and, from macro-economic point of view, only 10 per cent of all activities have a high degree of novelty. Therefore, 70 per cent from a micro-economic stance and 90 per cent from macro-economic point of view indicated only a middle to low degree of novelty in their innovation activities. Interpreting this result leads to the conclusion that companies more often develop innovations with a low or middle degree of novelty, regardless of the point of view.

In this work, the degree of novelty from the micro-economic point of view is used, as this view directly indicates the extent of influences on a company level, which has to be mentioned in strategic decisions. Therefore, a strategic decision has to consider the degree of novelty, especially because the higher the degree of novelty, the more changes have to be accepted in known elements (such as market needs, products of technologies) and therefore the more extensive the strategic decision has to be taken. To understand this relation between the degree of novelty and the extent of the changes to be made, different innovation types – incremental and radical innovations – are analyzed on different levels according to Kroy (1995: 63), including customer needs, markets, products and services, capabilities and basic science.

An incremental innovation represents a novelty that is based on existing products and services. Different parameters are modified or optimized on each level. For example, a subsequent technology is used to develop a new product or a product is changed in terms of new customer needs in a market. The connections between the different levels are therefore not detached, merely horizontally adjusted. However, radical innovations do not have all the vertical connections between the different levels from the outset. The innovation challenge of radical innovations is therefore primarily to develop the connections between the different levels. From each level an innovation could be triggered, whereas Leifer *et al.* (2000: 20) point out that 'the project [for radical innovations] often starts in R&D, migrates into some sort of incubating organization, and transitions into a goal driven project organization'. The project duration is also longer for radical innovations, with an average of ten years, then for incremental innovations, with an duration of six months to two years (Leifer *et al.*, 2000: 19). These differences in the project duration are mainly explained by the fact that, for incremental innovation, a more detailed plan can be developed which has relatively few uncertainties, while radical innovations are developed slowly during different less certain life cycles.[18]

Despite the uncertainties of radical innovations, which makes a decision more difficult, it is essential for companies to develop incremental as well as radical innovations, as Berth (2003) found through a survey. First, he found that radical innovations on average have a higher return than incremental innovation. However, much more interesting is a second point mentioned in the survey, that although the uncertainties are higher for radical innovations, the flop rate was the same as for incremental innovations. Therefore, one conclusion of the survey is that companies have to be more aware of radical innovations and include them more explicitly in strategic decisions.

Therefore, a strategic decision has to mention radical as well as incremental innovations, whereas in the case of radical innovation planning is more extensive. A decision on radical innovation has to be detailed more extensively from the beginning in order to connect Kroy's (1995) levels as soon as possible. For this reason, the strategic planning of radical as well as incremental innovation has to be considered on a strategic management level. A concept for formulating an innovation strategy has to consider incremental as well as radical innovations; therefore, the newness of an innovation has to be considered.

Object of the innovation

After the characterization of innovations by their degree of novelty in the last section, the innovations will now be categorized by the considered object, in terms of its concerns in innovation development activities. The term 'innovation' does not generally consider only the product-market field, but also process, management and organizational aspects. Therefore, the term 'innovation' is differentiated by the different aims. The boundaries between the different categories of innovations are now compared.

Thom (1980: 22ff.) proposes a differentiation of innovation into 'product innovation', 'process innovation' and 'social innovation'. Kaplaner (1986: 9) adds to this triple the term 'structural innovation'. Knight's (1967: 486) categorization encompasses near product-, process- and organizational focus together with aspects of the human resources area in the definition 'people innovation'. 'People innovation' comprises changes in company staff by acquisitions or dismissals, as well as the wanton change of behaviours. A classification into business, technological and organizational innovations is proposed by Zahn and Weidler (1995: 359). The authors cover a holistic and integrated approach with their definition of innovation (see Figure 2.7).

Zahn and Weidler (1995: 359) describe categories of innovation that show a differentiated categorization from the other authors cited. This is very clear in the integration of products and processes into one common innovation category, 'technological innovation'. Zahn and Weidler additionally integrate new – which was unappreciated until this concept – aspects, such as 'business innovation', and also 'culture'

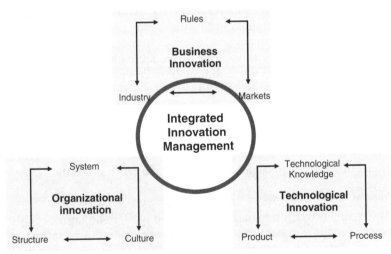

Source: Zahn and Weidler (1995: 359), reproduced with the permission of Schäffer-Poeschel Verlag fuer Wirtschaft Steuern recht GmbH Stuttgart

Figure 2.7 Integrated innovation: understanding innovation categories

and 'system'. There is also the explicit achievement of 'technological knowledge', which is the basis for developing technological competencies in a company.

The differentiation of innovation into these categories suggests that innovations appear individually. But, for example, product innovations in nearly every case appear with process innovations (Rammert, 1988: 199), so a new product often also demands a modification or renewal of processes. Utterback (1994: 217) proposes therefore that 'success ... requires equal emphasis on production and process design, which must be closely integrated'. In this context, Pfeiffer (1991: 43) found that companies possess excellent product innovations, but they do not actively profit in their competitive position. One reason for this, according to Wheelwright and Clark (1992: 73), could be that 'often [in] development projects, mean product development projects, the assumption being that process technology can be acquired easily if and when the need for it becomes obvious. Unfortunately, such a view results frequently in the full benefits of the product technology never being realized – the manufacturing process simply cannot deliver the quality, cost, or timeliness the product requires'.[19]

To conclude, it is necessary to divide innovations into categories to be aware of the possibilities, but in terms of realization, all the innovation categories have to be considered and developed together. This consideration of the innovation categories together is called 'integrated innovation'. This integrated innovation approach recovers any overlapping of business, organizational or technological innovations. In doing so, the resulting innovations have a broader range, additional protection against imitation and thus allow companies to gain sustainable competitive advantage by differentiating themselves (Zahn and Weidler, 1995: 359).

This integrated approach to innovations is an important element to consider in strategic decisions. Therefore, an innovation strategy formulation concept should allow business, technological and organizational innovations to be defined in terms of integral innovations.

Innovation management

The term 'innovation management' is used often in literature, but the term changes in relation to its tasks and responsibilities (Hoffmann-Ripken, 2003: 91). Therefore, the term 'innovation management' is now described in more detail.

Strategic management is described (p. 13) as having to design, direct and develop the company based on a clear understanding of systems complexity, systemic interactions and evolution. In the context of managing innovations, it is also necessary to understand the complexity, systemic interactions and evolution of the system. At this point, a question upraises: What is the exact system to be understood?

In this context, Brockhoff (1995: 986) distinguishes between an institutional and a functional view of innovation management. These two views are different in their system boundaries. The institutional innovation management says that the responsibility for innovation in a company is reduced to a group of people. This view is criticized because of the fact that innovation can be generated by every person in the company. Contrary to this view, the functional view can be taken into account, which is divided into the system theoretical view and the process view (Brockhoff, 1995: 987; Hauschildt, 1993: 23). The system theoretical view is concerned with the system in which innovations have to be developed and analyzed, such as the organization, the context and culture. The literature in this domain, which is concerned especially with organizational theory as well as cultural problems, is somewhat unmanageable. This is because the system theoretical view of innovation management is defined too broadly (Hoffmann-Ripken, 2003: 92). The process view of innovation management is primarily concerned with the decision about innovations, how innovations are implemented in the company and how innovations are developed. In this context, Hauschildt (1997: 25) defines the task of innovation management as how to develop the individual innovation processes in a anticipative[20] manner. These processes are the formulation of an innovation strategy and the organizational design of an innovation organization, including the processes utilized to realize the innovation intentions.

Based on the process view, an understanding of the innovation system is attempted: innovation processes, part of the innovation system, define values (Tipotsch, 1997: 55), and the processes of sales, production and the supply chain, as part of the delivery system, provide values (Tipotsch, 1997: 55).[21] Therefore, the 'innovation system' has the aim of defining values and the 'delivery system' has the aim of providing values. These two sub-systems together create new values, which leads to the concept of value creation by Porter (1985: 37) (see Figure 2.8). Therefore, the sub-system in a company that is concerned with innovation consists of value defining processes that conclude that innovation management has to design, direct and develop the innovation system based on a clear understanding of the complexity, systemic interaction and evolution innovation system.

36 *Structured Creativity*

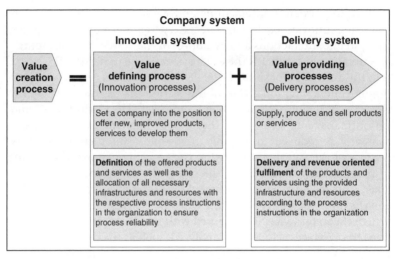

Figure 2.8 Value creation is the addition of defining value and providing value

At first glance, this description of innovation management could be seen as a complete definition, but there is still an open question to be answered before doing so: What elements have exactly to be understood in the innovation system in terms of complexity, systemic interaction and evolution? In this context, Afuah (1998: 14) and Bessant (2003: 6) argue that knowledge is the essential element to consider in an innovation system. Therefore, in the next section a more detailed discussion on knowledge is presented.

Knowledge[22]

According to Davis and Botkin (1994: 166), data comes to us in four different forms: numbers, words, sounds and images. They are worthless until they are related in a particular context. Information is analyzed data that has been arranged into meaningful, object-oriented patterns. The value and usefulness of information depends on the absorptive capacity[23] of the recipient (Cohen and Levinthal, 1990: 132). Once the recipient assimilates, interprets, evaluates and uses the information, we can talk about knowledge (Koruna, 2001: 100).[24] In this book, knowledge is understood as the totality of experience, cognition and skills of individuals to solve a problem.[25] Thus, knowledge is always action-oriented and personal. Speaking about the knowledge of the company, Probst, Raub and Romhardt (1999: 46) defines 'the organizational knowledge base as the totality of individual and collective experience, cognition and skills which the organization can access in order to solve a problem, including all underlying data and information'. Polanyi (1966: 4) argues that only a portion of knowledge can be articulated (explicit knowledge) and therefore be transferred. Implicit knowledge, on the other hand, can be neither easily articulated nor transferred via information or data, but can only be transferred through direct interaction between individuals via metaphors and analogies. Nonaka and Takeuchi (1995: 62) adopt this epistemology (theory of cognition) and describe in detail four modes of how knowledge can be converted: from tacit to tacit (socialization), from tacit to explicit (externalization), from explicit to explicit (combination) and from explicit to tacit (internalization). In addition, they argue that an iterative interaction between tacit and explicit knowledge (the knowledge spiral) becomes greater in scale as it moves up the ontological levels from an individual to a higher level.[26]

According to several authors,[27] knowledge is an overwhelmingly important productive resource in terms of market value and the primary source of competitive advantage. In this context, Warnecke (2003: 11) noted that the capability of managing this knowledge will be the decisive competitive factor in building the preconditions for economic success and to create innovations. Therefore, knowledge is the essential element to consider in the innovation system of a company. Afuah (1998: 14) argues that 'knowledge underpins a firm's ability to offer products, a change in knowledge implies a change in the firm's ability to offer a new product' and Bessant (2003: 6) noted that 'knowledge provides the fuel for innovations'. Therefore, 'a company which wants to innovate needs knowledge' (Bullinger, 2003: 261). These insights on the importance of knowledge in the innovation system draw the conclusion that innovation management has first to understand the complexity, systemic interaction and evolution of the knowledge in the innovation system (innovation relevant knowledge). This understanding of knowledge is a major challenge because knowledge is characterized by a high degree of complexity (Kuivalainen et al., 2003: 242; Schlaak, 1999: 42) and intransparency (Kroy, 1995: 77).

Conclusion

In sum, innovation management has to design, direct and develop the innovation system of the company based on a clear understanding of the complexity, systemic interaction and evolution of the innovation relevant knowledge. Furthermore, innovation management has to consider the strategic innovation-relevant criteria mentioned previously (pp. 26–37). These criteria are:

- innovation barriers (p. 26)
- innovation newness (p. 28)
- integral innovation (p. 32)
- innovation relevant knowledge (p. 36).

Innovation strategy and innovation strategy formulation

In literature, the term 'innovation strategy' is a rather new term and was rarely used in the past (Hoffmann-Ripken, 2003: 94). By virtue of the fact that innovation is considered important in the strategic context, on the business level as well as on the economic level, there is not much material published about innovation strategy (Gilbert, 1994: 16). The reason for this is probably, firstly, a deficiency of methodological concepts that integrates strategy into innovation (Olschowy, 1990: 32), and secondly, theoretically developed concepts of integrated innovation management have not been accepted in practice (cf. Thom, 1976).

In spite of the fact that the term 'innovation strategy' is not commonly used in literature, there are some indications that the subject has been developing in the last five years. Many recent authors consider an innovation strategy as important and provide definitions.

Innovation strategy

In the following, some definitions of innovation strategy are listed that differ in its extent. These definitions do not claim to be complete, but they are representative for the understanding in literature of the term 'innovation strategy' and other related terms:

- 'The innovation strategy is a description of the innovation process, which comprehends action oriented or/and action descriptions statements which are fitted to the organizational context of the innovation processes (objectives, members, structures, performances, boundaries and environment)':[28] Aregger (1976: 118)
- 'An innovation strategy tells us what actions a firm will take, when, and how it allocates its innovation resources': Afuah (1998: 99)
- 'An innovation strategy defines the long term objectives and fundamental directions of the innovation activities of the company and is therefore an integral part of the strategic fundamental direction of the company':[29] Schlegelmilch (1999: 106)

State of the Art in Theory 39

- 'An innovation strategy must cope with an external environment that is complex and ever-changing, with considerable uncertainties about present and future developments in technology, competitive threats and market (and non-market) demands': Tidd, Bessant and Pavitt (2001: 65)
- 'Patterns of activities about when and how to use new knowledge to offer products of services': Afuah (2002: 369)
- 'An innovation oriented strategy creates new products and processes as well as new forms of client interaction. The objective is not to pass the competitors on the same path, but to pass him in an innovative way to ensure the company's sustainable competitive advantage ... In the focus of the innovation oriented strategy are not only products, but also production processes and company structures have to be considered':[30] Bullinger and Auernhammer (2003: 29)
- 'An innovation strategy defines on a contextual level on the one hand the degree of newness and the direction of the aimed innovations. On the other hand the innovation strategy defines innovation type and its contribution to the company development. On a process level the innovation strategy is concerned firstly with the question of how innovation in the concept of the strategy can be integrated. Secondly, which factors the process affect that in their result will define the aimed conceptual as well as practicable application of the strategic induced and operative emergent different innovation types':[31] Hoffmann-Ripken (2003: 97)

It is obvious that these definitions of innovation strategy differ in part quite considerably. For example, Schlegelmilch (1999: 106) includes only innovation activities and, in contrast, Bullinger and Auernhammer (2003: 29) include the change of companies' structures as part of an innovation strategy. Because of this discord between the different authors, it is difficult to find an appropriate definition of an innovation strategy. Therefore, it is essential in a first step to know the content of a general strategy, and to know the particular context of innovation for the term 'strategy' and for 'innovation context'. Based on this, an innovation strategy should provide an understanding of direction, focus, organization and consistency as well as an innovation-specific understanding of integral innovations, innovation barriers, innovation newness and innovation-relevant knowledge. In Figure 2.9, these eight criteria are cross-matched with the definitions of innovation strategies.

40 Structured Creativity

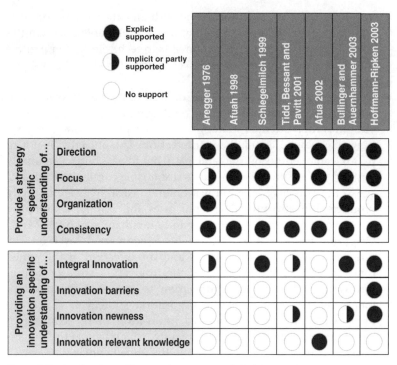

Figure 2.9 Evaluation of innovation strategy definitions

Figure 2.9 shows first of all that the understanding of the term 'innovation strategy' differs strongly from author to author. It stands out that Afuah (1998: 99) does not especially consider innovation-specific factors. On the other hand, authors such as Hoffmann-Ripken (2003: 97) consider very strong innovation-specific factors in their understanding of innovation strategy. A second interesting point is the trend that over time the term 'innovation strategy' has extended in its context. A third point is that there is one author, Afuah (2002: 369), who includes innovation-relevant knowledge in his understanding of innovation strategy.

Summarizing the different definitions of innovation strategy, it can be said that the addition of these definitions considers all strategy-specific and innovation-specific criteria. Therefore, the following definition of an innovation strategy is to be seen as a summary of the actual innovation literature and is based on the understanding of strategy (p. 13) and innovation management (p. 34):

An innovation strategy sets direction, focuses efforts, allows the design of an organization and ensures constancy in the innovation system while considering integral innovations, innovation barriers, and the degree of newness of the innovation as well as the required innovation relevant knowledge.

This attempt to define 'innovation strategy' is understood as underlying working definition and is used in this work.

Next, the term 'innovation strategy' will be analyzed from several points of view. In doing so, the authors wish to explain the different types of innovation strategy and also the formulation processes of an innovation strategy.

Types of innovation strategy

Neither in the innovation strategy nor in the general strategy research is there a unique understanding of the different innovation strategy types (Altmann, 2003: 44). The following overview shows different classification examples of innovation strategies in literature. It must be mentioned that some of the following classifications are not based on the term 'innovation strategy' but on related strategies, such as R&D strategy:

- Ansoff and Stewart (1967) (aggregation level: functional unit): first-to-market, follow the leader, application engineering, me-too
- Sherman (1982) (aggregation level: functional unit): first-to-market, me-too, technology push, market pull
- Zörgiebel (1983) (aggregation level: business unit): general technology leadership, general cost leadership, segment specific technology leadership, application specialization
- Porter (1985) (aggregation level: product): first-to-market (fast follower), late-to-market (cost minimization), market segmentation (specialist)
- Servatius (1985) (aggregation level: functional unit): the combination of:
    General strategy: focus, differentiate, standardize Time of the market introduction: active leader, passive follower, technology position, presence, leadership Technology: product, operating resources
- Cooper (1985) (aggregation level: company): technology induced strategy, balanced focus strategy, strategy with low technology risk,

strategy with low capital investment, high risk diversification strategy
- Abernathy and Clark (1985) (aggregation level: functional unit): functional unit technology: Hold or invest existing vs development of new technologies; functional unit marketing: hold or invest existing vs development of new markets
- Zahn (1986) (aggregation level: company, functional unit, business unit): pioneer strategy, imitation strategy, niche strategy, cooperation strategy
- Foxall and Johnston (1987) (aggregation level: product): market penetration, minor diversification (product), minor diversification (market), major diversification (product and market)
- Brockhoff and Chakrabarti (1988) (aggregation level: functional unit): aggressive specialist, aggressive innovator, defensive imitator, process developer.

This heterogeneity should attempt to reduce these dimensions. Additionally, this work focuses on the aggregation level of a product containing, on one hand, technologies and, on the other, offered products for a market. This work does not focus on the company, functional or business unit aggregation level; a simplified dichotomy categorization for an innovation strategy is chosen for this book:[32]

- Market leadership
- Technology leadership.

These two extreme dimensions of an innovation strategy are dependent, and they can therefore be developed in one company simultaneously. The challenge is to find the right company-specific balance between these two categories when formulating an innovation strategy.

Innovation strategy formulation

Up to this point, the understanding of innovation strategy in this work is clear. Nevertheless, there are different methods for formulating an innovation strategy. Five different approaches for formulating an innovation strategy are now presented[33] that could be seen as representative but not complete in the context of innovation strategy.

Quinn (1985) assumes that innovation can and must be integrated in the concept of corporate strategy. The strategy has to be defined according to him in which domain innovations are desired. With the aid of motivation and control systems, it is possible to identify the general conditions for encouraging the creativity and entrepreneurial potential of an organization and to develop it in a goal-oriented manner (Quinn, 1985: 80f.). Quinn recognizes that not all innovation activities can be planned, and advises a flexible strategic orientation to identify and react spontaneously to emerging opportunities. Nevertheless, it is important in the context of corporate strategy formulation to formulate an innovation strategy explicitly.

A similar approach for the formulation of an innovation strategy is presented by Martensen and Dahlgaard (1999).[34] The underlying understanding of strategy is that of the planning school. Innovation strategy is planned by the company top management in alignment with the vision and corporate strategy. Subsequently, acquisition plans and objective plans are developed, which are supported by an appropriate management commitment which is communicated to all employees in the company. In this process, the opportunities are evaluated with the aim of filling the innovation gap in the company. These results are part of the innovation strategy. An important point in this innovation strategy is the alignment of the objectives with the company-specific culture. Therefore, a loop is part of the formulation process which allows integrating feedbacks and the making of corrections. For the formulation of the strategy, it is essential that three questions are answered:[35]

- What are the firms' capabilities: Where are we and what can we do?
- What is wanted by the firm's customers: product/market pull?
- What is technologically possible for the firm: technology push?

Unplanned and operative emergent opportunities are not included in this concept.

Another perspective is the model proposed by Kawai (1992), who integrates strategic induced as well as emergent, understood as unplanned, innovations. Kawai starts from the view point, that innovations have to be anchored in the strategy. Kawai identifies two environmental conditions that need different innovation mechanism. If the environment does not have too many uncertainties and does not have too many structural changes, middle management can decide about the innovation activities. Should the environment be highly uncertain and many changes are appearing in the structure, it is necessary to pass through a strategic analytical formulation process at a top management level.

A fourth approach is presented by Tschirky (1998: 294f.; 2003: 58f.), which is more technology oriented. The author presents the formulation of a technology strategy that could be seen in the context of this work as a part of the innovation strategy. Tschirky sees the formulation of technology strategy as a special issue in the process of formulating corporate strategy, and parallel to other functional strategies (for example, the finance strategy). Tschirky describes a six-step process for formulating technology strategy: setting strategic objectives, analyzing the environment, analyzing the company, elaborating strategic options, taking strategic decisions and implementing the strategy (see Figure 2.10).

A further concept for formulating an innovation strategy, the so-called strategic innovation process, is presented by Afuah (1998; 2002). In this formulation process, a firm sets a mission and goals in a first step. 'To achieve these goals, it [the company] scans its environment and other sources of innovation for any opportunities and threats where it can exploit some of these opportunities and threats, the firm then chooses a profit site – that is, whether to be a supplier, manufacturer, complementary innovator, distributor, or customer. Next the firm formulates several strategies. In its business strategy, it decides whether the product – outcome oft the innovation process – will be low cost, differentiated, or both. In its innovation strategy, it decides whether to be the first to introduce the innovation or to be a follower of some kind. Business and innovation strategies help drive functional strategies – resource allocation and the actions taken by each function along the value chain (R&D, manufacturing, and so on) ... All these strategies constitute a strategic direction, which drives the implementation process that follows. Implementation entails having the right organization structure, systems or processes, and people' (Afuah, 1998: 335).

These concepts of innovation strategy formulation are obviously different; in the next section the five concepts are evaluated.

Evaluation and conclusion

The evaluation of the innovation strategy formulation concepts is based on the essential criteria in strategic management (p. 13), strategy formulation (p. 21) and innovation management (p. 34) presented in Figure 2.1. In Figure 2.11 the evaluation is shown and will now be discussed.

The authors Quinn and Kawai are aware of the fact that innovation is a special subject in relation to corporate strategy, but they do not consider the subject of innovation as special. Therefore, innovation-specific criteria have a tendency to be neglected. Martensen, Dahlgaard

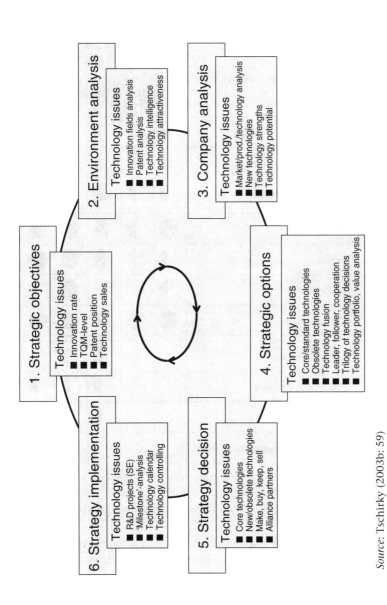

Source: Tschirky (2003b: 59)

Figure 2.10 Integration of technology issues into strategic business planning

46 *Structured Creativity*

		Quinn 1985	Kawai 1992	Tschirky 1998	Afuah 1998	Martensen and Dahlgaard 1999
● Explicit supported ◐ Implicit or partly supported ○ No support						
Provide a strategic management specific understanding of...	Understand complexity	◐	◐	◐	◐	◐
	Understand systemic interaction	◐	◐	◐	◐	◐
	Understand evolution	◐	◐	●	◐	◐
Provide a strategy specific understanding of...	Direction	●	◐	●	●	●
	Focus	◐	◐	●	●	●
	Organization	◐	○	○	●	◐
	Consistency	●	◐	●	●	●
Providing an innovation specific understanding of...	Integral innovation	○	○	◐*	●	○
	Innovation barriers	○	○	◐*	◐*	◐
	Innovation newness	○	○	◐*	○	○
	Innovation relevant knowledge	○	○	○	●	◐

*The authors point out the importance of this criteria, but not explicitly in the description of the formulation process

Figure 2.11 Evaluation of innovation strategy formulation concepts

and Tschirky also point out the importance of a special innovation-oriented strategy but, contrary to Quinn and Kawai, they mention in their strategy formulation concept the fact that innovation has specific criteria to consider. Despite this, Martensen and Dahlgaard focus only on products and Tschirky focuses only on technology, although Tschirky points out in another context that it is important to consider

State of the Art in Theory 47

integral innovations, innovation barriers and the newness of innovations. Afuah presents a concept of strategy formulation that considers innovation-specific criteria to be the best, whereas the consideration of the newness of innovation is not especially considered. All concepts do not explicitly consider the understanding of the innovation system in terms of complexity, systemic interaction and evolution. **This is the first gap in innovation strategy literature; the complexity, systemic interaction and evolution of the innovation system are not given special consideration in order to understand an innovation system so as to be able to formulate an innovation strategy.**

Beside the evaluation undertaken on the basis of the criteria by a more exact examination of the concepts, the concepts presented describe the major steps and criteria to consider but, with exception of Tschirky, they present no specific methods to use. This is a major lack in the literature, because these concepts cannot be taken into practice without investing effort. Only Tschirky's concept presents specific methods to use in practice, such as the technology portfolio and the technology oriented industry analyzes. Because Tschirky focuses mainly on technology aspects, his tool set is not a complete set for designing an innovation strategy formulation process. Therefore, **a second gap in literature is a missing structured and practitioner-oriented innovation strategy formulation process.**

At this point in the book, the gaps in the literature of innovation strategy formulation are clear. Therefore, in the next section solutions in complementary literature are sought to fill the gaps partly, or even completely. In doing so, the research in related literature focuses on the first gap in the literature, because the second (to build a practitioner-oriented innovation strategy formulation process including specific methods) is very dependent on the solution to the first gap. More specific changes in style and of the extension information of the model in order to understand the innovation system in terms of complexity, systemic interaction and evolution will directly influence the methods that can be used in the practitioner-oriented innovation strategy formulation.

Complementary literature

Focusing on the complementary literature in order to understand a system in terms of complexity, systemic interaction and evolution leads to a twofold procedure. As shown in Figure 2.3, strategic management has two tasks that differ in their time horizon; First to direct and design a company from 'today to tomorrow' (and therefore mainly to understand the existing potential for success and the actual system in terms of complexity and systemic interaction), and second, to develop a company from 'today to after tomorrow' (which requires a more specific understanding of future success positions in terms of the evolution of the future system). The next section gives an overview of concepts that allow the understanding of a system.

System models

To understand a system, one must be able to describe a system. This can be done, according to Whitehead (1997: 118), with a model. A model of a system is primarily a tool to communicate with clients, builders and users; they are the language of the system designer. 'Models enable, guide, and help assess the construction of systems as they are progressively developed and refined. After the system is built, models, from simulators to operating manuals, help describe and diagnose its operation.' Summarizing, a model is, according to Rechtin and Meier (1997: 13), an abstraction or representation of the system used to predict and analyze, in an example performance, costs, schedules, and risks, and to provide guidelines for systems research, development, design, manufacture and management. With this understanding, a model has the following roles, according to Whitehead (1997: 120):

- Communication with client users and builders
- Maintenance of system integrity through coordination of design activities
- Assisting design by providing templates, and organizing and recoring decisions
- Exploration and manipulation of solution parameters and characteristics; guiding and recording aggregation and decomposition of system functions, components, and objects
- Performance prediction; identification of critical system elements
- Providing acceptance criteria for certification for use.

All these roles in a model support the understanding of a system in terms of complexity and systemic interaction.

To identify the right model, it is necessary to know the objective of the intention to use a model. In the case of this work, it is essential to understand a system. Therefore the question 'What are the elements retained in the system and its interaction?' has to be answered, which implies a data model. Before analyzing the literature about models, the element to represent in the data model has to be identified. This is knowledge as discussed previously (p. 36). Therefore, in the following the literature that represents knowledge in the form of data models will be analyzed.

Data models in literature

There are several accepted data models of knowledge representation, according to Gordon (2000: 74). Some of these are presented in the following:

- **Rules** are reasonably easily understood by people and are also a powerful machine based on a knowledge representation scheme. They take the general form: If attribute A1 has value V1; and attribute A2 has value V2; attribute A3 has value V3.
- A **frame** is a collection of information and associated actions that represents a simple concept. It would be possible to represent a person (in a simple way by their name, date of birth and address) by the use of a frame. Frames can be used to represent complex pieces of knowledge and can also be constructed and edited as required.
- **Semantic networks** are easily understood by people and can be used in automated processing systems. This means that they can also become a vehicle to archive company knowledge. In a simple network, nodes are specific items and links show relationships between the items.
- **Concept diagrams** are closely related to semantic networks. Concept diagrams are also composed of nodes and arcs, and the nodes and arcs have similar functions. Concept diagrams can be used to describe fairly complex concepts. They are seen as knowledge-representational methods that employ graphical structures (Sowa, 1984).
- An **architecture** is a structure – in terms of components, connections, and constraints – of a system, according to Rechtin and Meier (1997: 253). The activity of creating an architecture is called architecting (see Figure 2.12). Architecting transforms problem and solution know-how into a new architecture. In doing so it is the objective to strive for fit, balance and compromise among the tensions of all the components.[36]

An architecture is not a detailed plan. It identifies the major components to be built, but does not specify exactly how they are to be built.[37] This is its major difference from semantic networks. For example, the

50 Structured Creativity

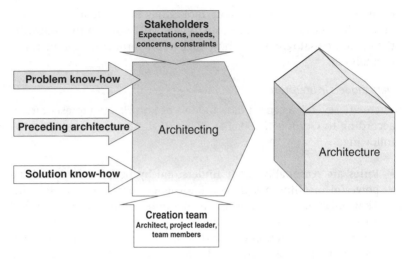

Figure 2.12 Architecting = creating an architecture

semantic network describes the effects of parts in a machine and how they interact. An architecture would show the machine as an overall system and its modules with their interaction, but does not show how they interact in detail. Therefore an architecture allows in the case of a complex and systemic interacting system – and to the disregard of the 'how' – the reduction of the complexity. This is a major advantage of architecture in the context of this book. For that reason, architecture seems to be the most interesting model for understanding a system in terms of complexity and systemic interaction.

The architecture as data model

After this brief introduction to architecture, it is necessary to get a sense of the possibilities of an architecture. There exist a multiple of different perspectives of architectures. Probably the most well known is the architecture of a building. Additionally, architectures are used for designing and structuring a product (product architecture), a software (software architecture) an organization (organizational architecture), information (information architecture) and even a strategy (strategic architecture). In the following, strategic architecture is described in greater detail.

The term 'strategic architecture' is mainly shaped by Hamel and Prahalad (1994: 107ff.). According to these authors, the term 'strategic architecture is basically a high-level blueprint for the deployment of new functionalities, the acquisition of new competencies or the migration of existing competencies, and the reconfiguration of the interface with customers ... A strategic architecture is not a detailed plan. It identifies the major capabilities to be built, but does not specify exactly how they are to be built ... A strategic architecture identifies what we must be doing right now to intercept the future. A strategic architecture is the essential link between today and tomorrow, between short term and long term. It shows the organization what competencies it must begin building right now, what new customer groups it must begin to understand right now, what new channels it should be exploring right now, what new development priorities it should be pursuing right now to intercept the future. Strategic architecture is a broad opportunity approach plan. The question addressed by a strategic architecture is not what we must do to maximize our revenues or share in an existing product market, but what must we do today, in terms of competence acquisition, to prepare ourselves to capture a significant share of the future revenues in an emerging opportunity arena ... A strategic architecture does not last forever. Sooner or later tomorrow becomes today, and yesterday's foresight becomes today's conventional wisdom.' A strategic architecture should always be developed in light of the criteria foresight, breadth, uniqueness, consensus and action ability. If these criteria are fulfiled, strategic architecture could be seen as the map. But the fuel to follow the route shown on the map is not yet identified. To summarize, 'an architect must be capable of dreaming of things not yet created – a cathedral where there is now only a dusty plain, or an elegant span across a chasm that hasn't yet been crossed. But an architect must also be capable of producing a blueprint for how to turn the dream into reality' (Hamel and Prahalad, 1994: 107ff.).

Summarizing strategic architecture would serve the following purposes:

- to outline the enterprise's strategic vision, conceived in terms of combinations of existing competencies and acquisition or development of new ones
- to reinforce a strategic focus on competencies, and ensure that all strategic decisions (such as investments, divestments, acquisitions and alliances) are consistent with the overriding imperative of maintaining and developing competencies
- to discipline organizational behaviour (especially business behaviour) by constructing a uniform frame of reference and implicit

decision rule; namely, 'Does this action or decision contribute to or detract from our core competence?'
- to assist with the alignment of culture, structure, human resource management and information management within the enterprise – again, by providing a uniform frame of reference as an instrument of integration and coordination
- to assist with organizational learning exchange (information flows focused on identifying business development opportunities) and promote strategic awareness throughout the enterprise
- to enable identification of deficiencies in skills and technology, which would limit the ability of the enterprise to build new competencies.

In all architectures, whether it is a building or a strategic architecture, a similar concept of architecting is used to solve context-specific problems by using a specific style of architecting. Although the concept of architecting is similar, the result, the specific architecture type, is different, as presented in Figure 2.13.

The concept of architecture is the same for all the contexts: it is about building a structure – in terms of components, connections, and constraints – of a system. Only the context-specific visualization of the architecture in terms of the type of architecture is changing. For example, a product architecture structures the product system and an organizational architecture structures the organizational system. Therefore, the concept of structuring is the same. But the visualization of architecture in the form of context-specific architecture is not at all

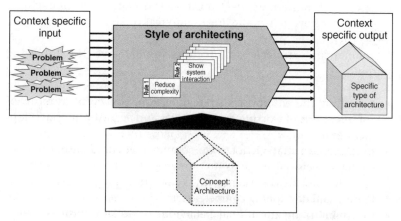

Figure 2.13 The context-specific use of the architecture concept

State of the Art in Theory 53

the same. So, the type of the architecture changes according to the context.

The style of architecting is different. Each architect is using a different style, which is defined by the rules to be used. These rules of architecting are often the same in all the specific contexts, as can be seen in Figure 2.14. More concretely, there are seven rules of architecture that are used in all the presented types of architecture: 'reduce complexity', 'describe systemic interaction', 'define system objectives', 'define system elements', 'visualize concept', 'define system to be model', 'ensure system consistency' and 'define operational interfaces'. This appearance of the seven main rules in every type of architecture is an indication that these rules are independent of the style of architecture. There are some rules, however, that only play a role in a few types of architecture. An example is the rule 'identify knowledge gap', which is used only in the context of 'product architecture', 'software architecture' and 'strategic architecture'. These rules are additional context specific rules. In contrast to the main rules, the authors do not pretend that the context-specific rules are all named. These findings, based on the analyses of the existing architectures, indicate that there are main and context-specific rules of architecting that allow a more detailed definition of the concept of an architecture:

> The **concept of an architecture** is a procedure to structure a system by its components, connections and constraints to fulfil seven main rules ('reduce complexity', 'show system interrelations', 'define system objectives', 'define system elements', 'visualize concept', 'define system to be model', 'ensure system consistency' and 'define operational interfaces') independent from the specific context.

In contrast, the definition of the type of architecture is:

> The **type of architecture** is a context-specific visualized solution of the concept of architecture to solve context-specific problems by using a specific style of architecting which is defined, on the one hand, by the main rules and, on the other hand, by the context-specific rules.

Conclusion

The concept of architecture corresponds to the need to understand an actual innovation system in terms of complexity and systemic interaction. Additionally, architectures allow more functionalities than

		Types of architecture					
		Building architecture	Product architecture	Software architecture	Organizational architecture	Information architecture	Strategic architecture
Main rules	Rule Reduce complexity	X	X	X	X	X	X
	Rule Describe systemic interactions	X	X	X	X	X	X
	Rule Define system objectives	X	X	X	X	X	X
	Rule Visualize concept	X	X	X	X	X	X
	Rule Define system elements	X	X	X	X	X	X
	Rule Define operational interfaces	X	X	X	X	X	X
	Rule Define system to be model	X	X	X	X	X	X
	Rule Ensure system consistency	X	X	X	X	X	X
Context specific rules	Rule Identify knowledge gap		X	X	X		X
	Rule Assign task			X	X	X	
	Rule Define processes/flows		(X)	X	(X)	X	
	Rule Plan resources				X		X
	Rule Embed system in a holistic context	X			X		X

Figure 2.14 The types of architecture

reducing complexity and describing the systemic interaction, as shown in Figure 2.14. In this context, Schaad's (2001) innovation architecture is a promising basis for developing an architecture representing the complexity and systemic interaction of an innovation system. However, there is still an innovation system-specific gap in this literature: **a specific type of architecture for understanding the complexity and system interaction by visualizing that the innovation relevant knowledge is still missing.** This is a gap to be closed.

In the domain of business intelligence, an extensive number of different methods to identify potential innovations can be found. This can be done with quantitative methods or with qualitative methods, such as functional, innovation field and core competencies-based methods. These methods are sufficient to identify potential innovations. However, no concept was found summarizing all these identified potential innovations to give a complete understanding of the evolution of the future system. Prahalad's and Hamel's (1990) concept of strategic architecture is a basis for developing such a model for the future. **This lack of a model to understand the evolution of the future system is a further gap in literature that should be closed.**

Conclusion: innovation strategy formulation (theory)

Summarizing the state of the art in the literature for innovation strategy formulation, the first main gap (p. 47) identified is:

> Complexity, systemic interaction and evolution of the innovation system are not adequately considered in order to understand an innovation system and to be able to formulate an innovation strategy.

Trying to close this gap, a more detailed research in complementary literature showed that the innovation architecture from Schaad (2001) is a basis for understanding the complexity and systemic interaction of the actual innovation system. Nevertheless, there is a more detailed gap (p. 47) remaining in the domain of architecting:

> A specific type of architecture in order to understand the complexity and system interaction by visualizing the innovation-relevant knowledge is missing.

Quantitative and qualitative methods presented in the literature are the basis for identifying new success potentials in order to understand the evolution. But also in this domain a gap (p. 55) was identified:

> There is a lack of a model to enable the understanding of the evolution of the future system.

A second main gap identified (p. 47) is:

> A structured practitioner-oriented innovation strategy formulation process is missing.

3
State of the Art in Practice

The previous chapter showed the importance of research in innovation strategy formulation from a theoretical point of view. The aim of this chapter is to clarify the relevance of the topic in practice. It is not the goal to examine practices in companies, but to examine interest in the topic. The practitioner's voice is captured by means of interviews conducted at the beginning of an action research project, conducted especially to identify requirements in the domain of innovation strategy in practice. The results of these interviews are summarized and the chapter closes with a conclusion of innovation strategy formulation from a practical point of view.

Interviews

Interviews were held in innovation driven companies. These interviews were conducted specifically to obtain answers to the following questions:

- Is the need for an innovation strategy and its formulation process a real need?
- Are there any concepts already in place that explicitly formulate an innovation strategy?
- Are there are concepts for formulating an innovation strategy? What is the concept in terms of best practice in industry?
- If there are no concepts for formulating an innovation strategy, what elements could be interesting for the process of innovation strategy formulation?

Based on these questions the interviewees were selected to be heterogeneous in terms of their company size and industry. These selected companies are presented in Figure 1.3. The company names have been changed because the interviewees were more open to answering questions in the knowledge that the specific company name would not be published.

The following persons were interviewees in different companies:

- **RubTec** (6 interviews): CTO, Research and Development General Manager and 4 Section Managers.
- **Toll Revenue** (2 interviews): Head of Innovation and Marketing, Product and Business Development Manager.
- **Optic Dye** (2 interviews): CEO, Director R&D.
- **TecChem** (10 interviews): CTO, 9 Segment R&D Managers.
- **HighTec** (5 interviews): General Manager (Technology and Innovation), Vice President Corporate Development, R&D Manager, Division CTO, Product Manager
- **Info Exchange** (2 interviews): Head of Development of IT-Infrastructure, Head of New Business Solutions.
- **StockTec** (2 interviews): CEO, Head of R&D
- **MicroSys** (2 interviews): R&D Manager, Manager of Advanced Technology
- **Built-up** (1 interview): Corporate Technology Manager.

Additionally, the subject of innovation strategy formulation and its lack in practice was discussed in the ERFA conference. In this discussion, eight representatives of the practice participated.

This sample of 32 interviews and the ERFA conference discussion made it possible to gain insight into the domain of innovation strategy formulation in practice. A major insight was that most companies have problems focusing their innovation activities in a single direction for effective and efficient innovation. Some companies already focus their activities significantly by integrating the subject of innovation into the corporate or strategic business unit strategy. Nevertheless, in these companies innovation aspects are not well considered on a strategic level, according to their statements. Most companies complain about difficulties in diverse stages in the strategy process that considers innovation. To summarize, the most important problems and needs are:

Importance of innovation strategy: Companies insist that a clear definition of the innovation intentions through an innovation strategy and seen as part of the corporate or business unit strategy, makes it possible to steer and align the activities in the innovation processes consistently through:

- a clear prioritization and synchronization of activities
- an effective and efficient deployment of resources (such as infrastructure, rights, patents, financial capital and knowledge)
- a clear definition of responsibilities
- a consideration of integrated innovations considering business, technology and organizational innovations
- a strategic control of innovation activities through a comparison of the clearly defined aims in the strategy and the results.

An additionally important point mentioned in interviews is that during the formulation process of an innovation strategy, the management team should be forced to think through all innovation opportunities and to evaluate them. This includes thinking about future required competencies in order to ensure the success of the innovation and to ensure the future success of the company.

Lack of an innovation strategy formulation process: All of the companies confirmed that they do consider innovation activities in the corporate or business unit strategy formulation process. All the companies confirmed that their innovation activities are essential to gain competitive advantage for the future. However, practically no strategy was found where concrete innovation activities could be derived and there was no structured strategy formulation process evident that considered innovation activities in an integrated manner. In general, the top management and middle management of R&D gather the necessary information about innovation activities based on specific meetings in order to update one another on their intentions. Unfortunately, this is not coordinated in a holistic and structured way, including marketing, development and research activities. The reasons given for this lack of structured innovation strategy formulation process are multifaceted:

- Innovation activities are very difficult to understand holistically and thus do not permit the formulation process to be undertaken in a structured manner.
- Development activities are very dynamic; therefore a decision today can soon be obsolete. At the same time, it was often mentioned that decisions are only obsolete because the decision was not well grounded.
- Some companies mentioned a lack of tools to support the building up of an innovation strategy. Other companies mentioned that they have a strategy process and strategic methods (such as portfolios, roadmaps and investment calculations) to prepare decisions but

these methods are very difficult to use in the domain of innovation because structured information about the company's innovation activities is not available.

This discussion has shown that companies are aware of the importance of innovation strategy. Furthermore, some efforts are made to integrate innovation activities further into strategy processes, but not at the required level. A pattern, however, could not be identified for innovation strategy formulation. These findings show that an innovation strategy is important but, at the same time, the absence of a formulation process is evident in most of the companies, which is in alignment with the study done by Kambil (2002: 8).

Beside the explicit requirement of an innovation strategy formulation process, it must be mentioned that companies do not have a need for a complete new innovation strategy process, but rather for a process that is integrated into the actual corporate or business unit strategy processes. Based on the statements from practice, a main focus should lie on structuring the information on innovation activities and intentions so as to integrate them into the existing strategy process with existing tools.

Conclusion: innovation strategy formulation (practice)

The aim of this chapter is to gain insight from practitioners about problems and needs of innovation driven companies in relation to innovation strategy formulation. Indeed, innovation strategy formulation is of concern in innovation driven companies' reality. Several interviews (in 11 companies) showed that innovation strategy formulation is of serious interest to innovation driven companies. Apparently, no systematically concepts of innovation strategy formulation could be found. Therefore, one conclusion is:

> Most companies agree on the importance of an innovation strategy. However, they had no explicit innovation strategy formulation process. Therefore, there is a **call from reality** for a structured, practitioner-oriented innovation strategy formulation process for innovation driven companies.

In the next chapter this call from practice will be compared with the gap in the literature in order to define the focus and requirements of a solution concept.

4
The Dual Gap

This chapter is the bridging link between fulfilled and unfulfilled research in the field of innovation strategy formulation in innovation driven companies. Thus, the aim of this chapter is briefly to describe the dual gap based on the gaps identified during the study of theory and practical reality, to summarize the requirements of the solution concept and to present working hypotheses.

Dual gap in innovation strategy formulation

The previous chapters showed a gap in the research of innovation strategy formulation in innovation driven companies from a theoretical and a practical point of view. The practitioner's voice demands a practitioner-oriented strategy formulation process. However, the literature in innovation strategy formulation does not present such a process, and complementary literature has basic concepts, such as innovation architecture, that are promising. This leads to the following two dual gaps, discovered in literature as well as in practice, which have to be closed in this book:

> **First dual gap**: a concept for understanding the complexity, systemic interaction and evolution of the innovation system is missing.

> **Second dual gap**: a structured, practitioner-oriented strategy formulation process and its implementation are missing.

To close these gaps, it is first of all necessary to define the criteria for a possible solution. These criteria were developed in the form of evaluation criteria in the section on the state of the art in literature and are now summarized:

- The innovation strategy formulation process should provide an understanding of the complexity, systemic interaction and evolution of the innovation system (p. 41)
- The innovation strategy formulation process should provide an understanding of focus on direction, organization and consistency for the innovation system (p. 41)
- The innovation strategy formulation process should provide an understanding of innovation-relevant knowledge, integral innovations, innovation barriers and the innovation newness (p. 41).

At the end of this book, these criteria will be included with the feedback from practice to provide an evaluation of the solution of the innovation strategy formulation process developed in this book.

Working hypotheses

Based on these gaps and insights, gained from theory and practice, one can formulate working hypotheses, which will be developed in this book. These working hypotheses are not understood as tentative assumptions that have to be tested, but as guiding ideas of this work to find answers to the research questions. Thus, the working hypotheses comply with the three research questions.

QUESTION 1:

How can a **complex, systemic interactive and evolutionary innovation system** be modelled that can **understand** the system specific conditions of an innovation driven enterprise?

WORKING HYPOTHESIS 1:

The concept of architecture is a solution for understanding the complex, systemic interactive and evolutionary system of innovation driven enterprises.

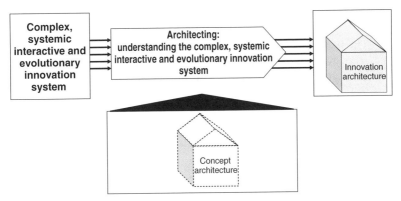

Figure 4.1 Working hypothesis 1: understanding the complex systemic interactive and evolutionary system

QUESTION 2:

How could a structured innovation strategy formulation concept be **designed**, based on the innovation architecture, for innovation driven enterprises?

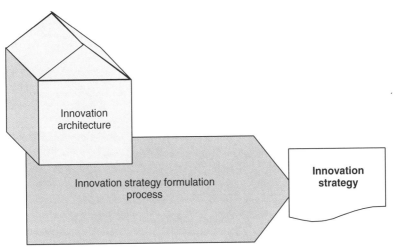

Figure 4.2 Working hypothesis 2: innovation architecture as innovation strategy formulation support

WORKING HYPOTHESIS 2:

The Innovation architecture applied in an adapted innovation strategy formulation process is a support for innovation driven companies to define an innovation strategy.

QUESTION 3:

How could such an innovation strategy formulation concept be **implemented**?

WORKING HYPOTHESIS 3:

Implementing an innovation strategy formulation process in an innovation driven company cannot be realized by implementing the whole theoretical process. However, the theoretical innovation strategy formulation process is a basis for adding the missing steps by which the company will be able to define an appropriate innovation strategy.

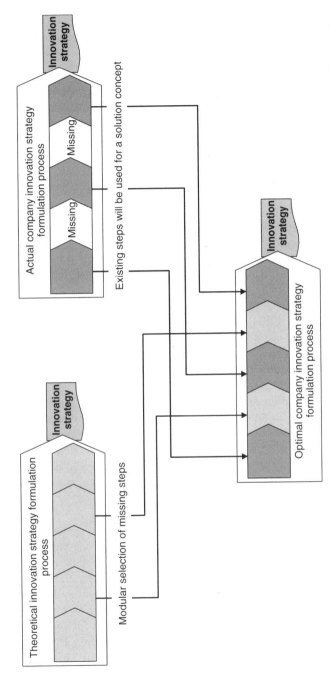

Figure 4.3 Working hypothesis 3: implementation of innovation strategy formulation process; company specific implementation

5
Concept

In the last chapter the dual gap identified in the literature and practice, as well as the state of the art in closing this gap, was shown. It was demonstrated that the state of the art does not offer solutions that would close these gaps. Therefore, in this chapter a concept is presented to close the gap by answering the three research questions in Chapter 4

The first research question – modelling a complex, systemic interactive and evolutionary innovation system in order to understand the system-specific conditions of an innovation driven enterprise – is answered with the concept of innovation architecture. The second and third research questions – to design and implement an innovation strategy formulation process – are also presented. The two concepts are presented in different chapter. Innovation architecture also plays a major role in Chapter 5 because the innovation strategy formulation process is based on this architecture

The concepts of innovation architecture and innovation strategy formulation must both consider certain innovation specific aspects as presented in the theoretical part of this book. In brief, the most important aspects are that:

- the complexity and systemic interaction of the system has to be modelled to understand the actual innovation system. The evolution has to be visualized in order to improve the sense of the future development of the innovation system (see chapter 2.1.1).
- the architecture as well as the innovation strategy formulation process should allow innovation opportunities to be evaluated and consistent decisions to be taken with a clear and focused direction. Additionally, the innovation strategy should allow sufficient organization for the innovation system to be derived (see chapter 2.1.2).

- the innovation-relevant knowledge has to be represented in the architecture in order to identify the existing knowledge as well as the required knowledge for the development of certain innovation opportunities. This is essential for developing an innovation strategy. Furthermore, all kinds of innovation should be considered in terms of newness and the purpose itself. Additional attention should be paid to innovation barriers (see Chapter 2).

These aspects are the basis for the development of the following concepts; innovation architecture and the innovation strategy formulation process.

Innovation architecture

Introduction

The solution to the problem of modelling a complex, systemic interactive and evolutionary innovation system is the architecture, the underlying concept as described in the first working hypothesis (see chapter 4.2). More specifically Schaad's (2001: 116) innovation architecture was the basis for designing a solution for an innovation architecture (Figure 5.1). This innovation architecture was redesigned for the specific requirements of this work and will be described in detail in the following sections.

This innovation architecture is a three-dimensional house of an innovation system analogous to the architecture of a building. In contrast to the architecture of a building, the three dimensions of an innovation system are not spanned by the three steric directions but by three knowledge dimensions: object knowledge, methodological knowledge and meta-knowledge (see Figure 5.2).

These three dimensions of knowledge are based on knowledge understanding in the domain of semantics, as defined by Wagner (2002). Wagner argues that when people communicate, they want to transfer knowledge. In doing so, people have the intention of transferring knowledge about objects (object knowledge), about actions (methodological knowledge), and about backgrounds of the transferred knowledge (meta-knowledge). Wagner argues that these three knowledge groups categorize all the knowledge that people want to communicate. So it can be said that this regrouping can represent the essential knowledge of a system.

More specific object knowledge is the knowledge about objects and information in our environment. In innovation, the object knowledge

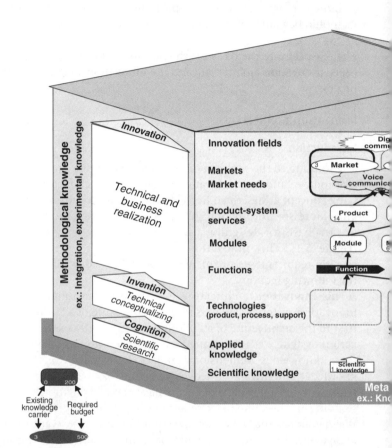

Figure 5.1 Innovation architecture

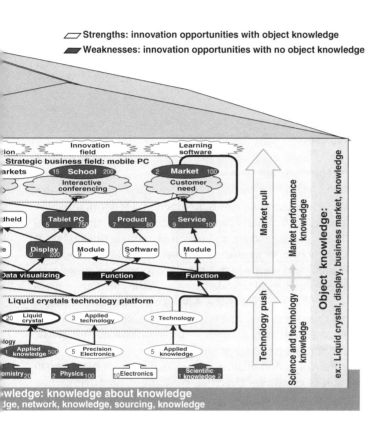

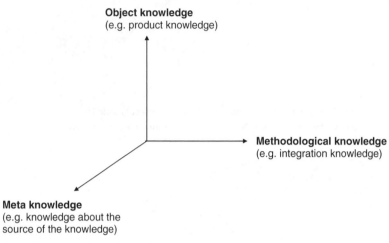

Figure 5.2 Three knowledge dimensions

is the knowledge about customer needs, products and services, modules, technologies and scientific insights applied. In contrast, methodological knowledge is the knowledge to act and to comply with actions. Thus, it is the knowledge about how to proceed and behave, what specific steps and tasks to initiate and with which procedure to complete the tasks. Especially, the methodological knowledge enables the creation, processing and transfer of object knowledge. Meta-knowledge encompasses knowledge about the source, reliability, importance, and transferability of the knowledge as well as the cognitive capabilities available to the knowledge development (cf. Wagner, 2002).

Combining the three dimensions of knowledge, the object knowledge is the result of creating, extracting, combining, modifying, integrating, modelling, applying, storing and transferring of more general object knowledge by using methodological knowledge. Metaphorically speaking, this understanding assumes an object knowledge flow that is similar to the material flow in production, where specific stages are stated in materials, parts, components, assemblies and finished goods. Between the stages, there are varying activities that create process and transfer the objects from one stage to the next. Along the material flow, the division of work is represented by the logical flow of activities that are designed into the product (almost inherently) during its conceptual design, development and/or industrialization. Similarly, the object knowledge (e.g. products, technologies) is processed during a logical flow of activities towards its innovative purpose and represented by stages that are identifiable along its processing. Specific methodological knowledge (e.g. analytical techniques or integration know-how) is required for its creation, processing and transfer at each stage, guided by specific meta-knowledge (e.g. scientific reasoning, modelling and transferability).

This three-dimensional knowledge concept is essential for the structure of innovation architecture. Each dimension is described in detail in the next three sections.

Object knowledge

In the dimension of object knowledge, object knowledge and its systemic interaction are visualized. The objects are markets, products or services, modules, technologies, applied knowledge and scientific knowledge. The structure of these objects is based on the principles of knowledge depth (see Figure 5.3).

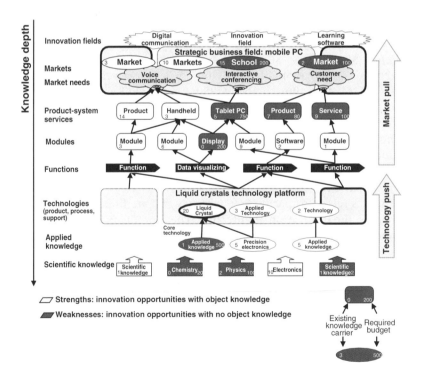

Figure 5.3 Object knowledge dimension

According to Schaad (2001: 109f), Knowledge depth is an indicator of how far a company advances its innovation efforts in the direction of new domains in order to make them accessible and to integrate them into concrete products. The entity of knowledge depth is degree of generalization or specialization of the object knowledge from a company's point of view. A high degree of generalization, and therefore a great degree of knowledge depth, means that the object knowledge is more general, but is not directly linked with the development of a specific product. This is typically scientific knowledge. Whereas, a high degree of specification of the object knowledge, and therefore a low degree of knowledge depth, means that the object knowledge is very specific and has a direct link to the development of a particular product by the company. Based on this knowledge depth, the objects are vertically structured as shown in Figure 5.3 and now described in greater detail.

The object knowledge of **markets and market needs** is the most specific type of object knowledge, because it is the knowledge about the requirements of an actual or potential customer. Typically, it is the knowledge that is developed in marketing. These markets and market needs are regrouped in innovation architecture by **strategic business fields (SBFs)**. These SBFs represent an isolated functioning part of company's market oriented activities (Müeller-Stewens and Lechner, 2001: 115). According to Abell (1980), such an SBF should be defined based on a three dimensional raster: the function describes the customer need, the customer groups describe the specific buying attitude, and the technologies describe the technical solution to fulfil a specific function.

Products are final goods and services that are offered by a firm based on utilizing the competencies that it possesses (cf. Teece, 1997). The products relate to specific market needs and are linked directly with them in the innovation architecture. These products are divided into **modules**, which are uncoupled parts of products. These parts should be designed in terms of respecting the rules of modularity. According to Mikkola and Gassmann (2003: 407f.), modularity refers to the scheme by which interfaces shared among components in a given product architecture are specified and standardized to allow for greater substitutability of components across product families. This increases the product variety and customization. When interfaces of components or modules within a system becomes standardized, outsourcing decisions can be made accordingly with respect to a firm's long-term strategic planning of its new product development, manufacturing and supply chain management activities. Additionally, modularization is an approach for organizing complex products and processes efficiently (Baldwin and Clark, 1997), by breaking down complex tasks into simpler portions so they can be managed independently. Modularization permits components to be produced separately, or loosely coupled (Orten and Weick, 1990; Sanchez and Mahoney, 1996), and used interchangeably in different configurations without compromising system integrity (Flamm, 1988; Garud and Kumaraswamy, 1993; Garud and Kumaraswamy, 1995). Therefore, in innovation architecture modules make sense in principle, in the event that the product can be divided into modules.

Technology is specific individual and collective knowledge in explicit and implicit forms for product and process-oriented usage based on natural, social and engineering-scientific knowledge. These technologies can be divided into product technologies and process technologies. **Product technologies** deploy scientific or engineering principles, dealing with a specific effect and determine how an effect occurs. This effect allows the fulfilment of a specific product function which, from the point of view of the market, is towards expected customer needs. For example 'liquid crystal technology' is a product technology for fulfilling the product function 'visualize data'. **'Process technologies** however, deploy the effects of an existing product technology. R&D process technologies are used to perform R&D activities and may include technologies such as microscopy, nano and atomic technology. Typical production process technologies include casting, milling, galvanizing, soldering and surface mounted technology. They also consist of logistics and quality assurance technologies. Administrative process technologies usually comprise office automation technologies and, finally, infrastructural process technologies typically may comprise security, elevator and air conditioning technologies. These different technologies can be regrouped in **strategic technology platforms (STPs)** which, as a structure, can be used to reduce the complexity of the usually large number of technologies to be handled. Such a technology platform, in literature also called the technology field, is a relatively isolated part of the actual and future technological activity field (Ewald, 1989; Peiffer, 1992: 65). STPs are the counterpart to strategic business fields (SBFs) (cf. Brodbeck, 1999: 22). The definition of an STP is based on the three dimensions: technology, theory and know-how (Brodbeck, 1999: 22ff.).

Applied knowledge and scientific knowledge are the objects with the highest knowledge depth in innovation architecture. They are both the basis for developing new technologies and, therefore, new products. But they are different on one essential point. **Applied knowledge** is the result of applied research which is, according to Picot (1988), the application-oriented search for knowledge for existing problems. In contrast, **scientific knowledge** is the result of basic research which is mainly concerned with an application neutral search of basically new knowledge (Picot, 1988).

With innovation architecture presented in Figure 5.3, it is possible to visualize, and therefore to understand, the innovation system. The innovation architecture visualizes the systemic interactions between the object knowledge. In doing so, not only is the existing innovation relevant knowledge integrated (white objects) but also the knowledge that must be developed for realizing a concrete innovation opportunity (grey objects). In other words, the knowledge that is already in-house is a strength, and the weaknesses are the knowledge not on the required level, and where in future innovation activities have to be focused.

Functions and innovation fields are also integrated, in the innovation architecture, in addition to the linked object knowledge. The **functions** are integrated into the object knowledge dimension for several reasons. Firstly, they allow a solution neutral connection between technologies and products/modules, which helps in the strategic steering that coordinates the solution oriented product development on an abstract level and the effect-oriented technology development. This makes it possible to ensure the handshake between **technology push** and **market pull**. Secondly, the functions allow new business fields and technology platforms to be identified and integrated, which give an overview of potential opportunities. Thirdly, the definition of the function, which is a strategic decision, gives the direction of future activities. For example, companies that produce microwaves can define a function as 'heat meal', which would define a strategic direction in the food business, which is not only limited to microwave technology. But it is also imaginable that the company would decide to define the function as 'rotate dipole', which is basically the technological effect of microwave technology. This function would allow a strategic direction that is much broader than for the food business alone. Therefore, the function is a central element in innovation architecture.

Because functions are so central to innovation management, especially in innovation architecture, the theory of functions is now presented.

A function is a solution neutral description of an operation that describes the constraints between input and output variables (Meier, 2002: 6). According to Biedermann (2002: 39f), a function describes

'What a product really does.' To analyze what a product really does, a function is often visualized as a black box which comes from the system techniques (Daenzer and Haberfellner, 1999); the specific material, energy of signal transformations result of a causal difference between the attributes of the input and output values (Meier, 2002: 6). The function is verbal, described with an input value consisting of a subject that is broadly composed and an operation in form of a verb that is very specific, such as 'transmit power'. However, it is still complicated to define a function with this basic rule because the levels of abstraction and detail have to be defined as appropriate to the system.

- Level of abstraction: The VDI (1996: 5) proposes in this context that a function should not be formulated in the real domain (Figure 5.4) because this would be only a verbal description of the reality containing the solution. Instead, a function should be described as being between the frontier of the iconic and symbolic domain. A

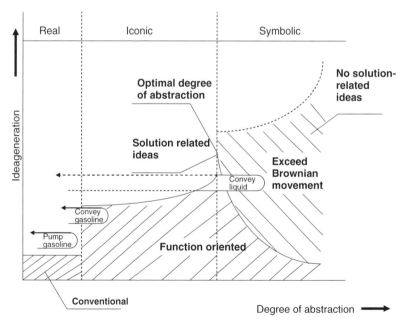

Source: VDI (1996: 5), reproduced with permission

Figure 5.4 Two-level progression of idea generation

settlement to fare in the symbolic domain would be too abstract to define the function.
- Level of detail: According VDI (1996: 3ff.), a function can be defined on different levels of detail as illustrated in Figure 5.5. The main function describes what the considered product really does; such as, a hammer drill removes material. This hammer drill can only work if the main function is supported by additional functions, such as supply energy. These two categories can be summarized in the main functions, which represent at the same time the 'as-is' situation of the product and the 'to-be' objectives of the product. This implies that this level of detail does not predefine conceptual solutions. The basic functions can be divided into detailed functions. For example, detailed functions of 'remove material' are 'disassociate material', 'scale down material' and 'take out material'. With these detailed functions, the solution to use a drill is already predefined. As can be seen, the level of detail for defining a function has to be appropriated to the needs. If it is necessary to rethink the whole concept of a product, the level should not be too detailed. In contrast, if an element of the product has to be redesigned, the functions should be more detailed.

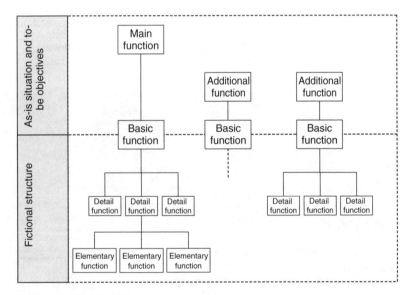

Figure 5.5   Functional tree

Defining a function with these fundamental guidelines leads to an abstract description which is of major benefit (Pahl and Beitz, 1993: 1). Such benefit is the support in the following domains, according to Meier (2002: 5f):

- To avoid a pre-fixation and fixed pattern of thought
- To structure an unstructured overall function in easy sub-functions
- To conjunct a solution with its effect
- To define a direction sign for allocating solutions
- To identify and define priorizations in development
- To minimize the effort in realization (minimal structure principal)
- To make acceptance analyses in the markets by functional value analyses
- To identify more solutions in the process of finding solutions.

As seen, the functions can be used for multifaceted tasks, where the final task is to identify new solutions, which could be seen as a way to identify new potential new successes. Therefore, the identification of new success positions based on the functional method are the focus of the next section.

Such a functional approach allows new business fields to be identified through the analysis of unfamiliar markets where this function could satisfy customer's needs. Also the approach allows new emerging technologies in unfamiliar technology platforms to be identified that fulfil the same function (cf. Pfeiffer *et al.*, 1997: 71f.). For example, the function of a photo camera is to store a visual image. To identify new business fields outside the photo industry, the question to be asked is: 'Who else needs to store visual images?' The answer could be that people who want to copy something also need this function, so the new business field could be in the area of copying machines. To identify new technology fields, the question must be asked: 'Which other technologies store visual images?' If, in the past, the technology was based on photochemical technologies, new technology and digital technologies now fulfil this function. This process helps to identify major changes in the environment in a timely manner.

To support the process of defining functions in innovation architecture, three major guidelines are essential:

1 **The definition of the functions is related to the product properties.** For products that are process oriented, the functions represent the different stations of the process flow. Where a product is not

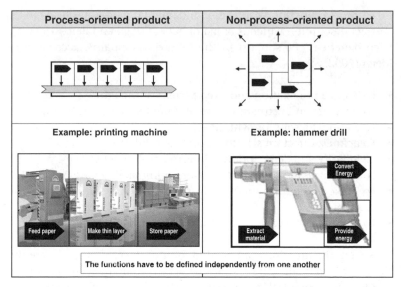

Figure 5.6 The definition of the functions is related to the product properties

process-oriented, the functions represent the modules of the product (see Figure 5.6).

2 **The definition of the functions relates to the technology decision making perspective.** Where innovation architecture is made for a normative decision perspective, the functions are more general, abstract and solution-oriented. Where innovation architecture is made for an operational decision perspective, as a steering instrument, the functions are more detailed, figurative and solution-oriented (see Figure 5.7).

3 **The definition of the functions relates to companies strategic intentions.** Where a company is highly market-oriented and wants to fulfil one specific customer need, the function has to express this strategic intention of fulfilling one customer need. Where a company is highly technology-oriented and wants to enter into several markets with different customer needs and with one single technology, the function has to express what the technology really does (see Figure 5.8).

Innovation fields have two main tasks; they allow new, but related, strategic business fields and technology platforms to be identified, and they are a steering instrument for defining the search fields for

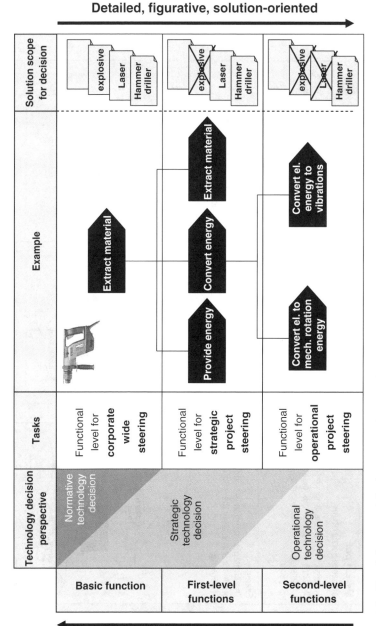

Figure 5.7 The definition of the functions relates to the technology decision-making perspective

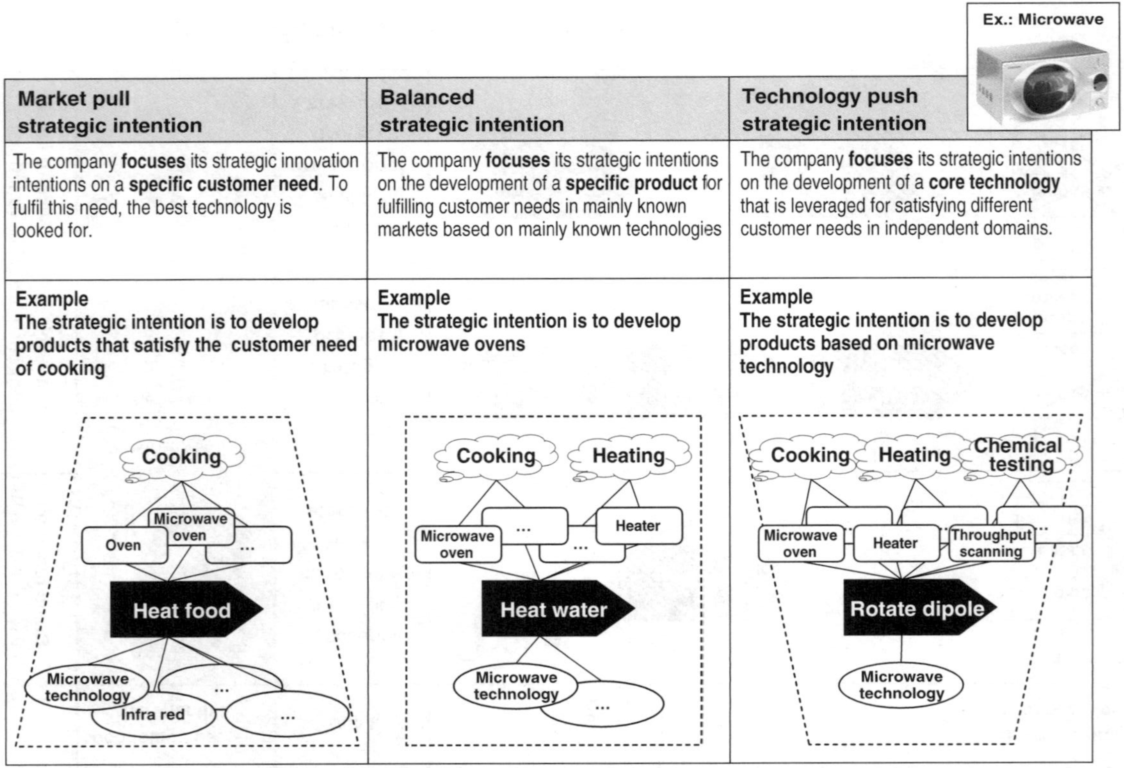
Ex.: Microwave
Market pull strategic intention
Balanced strategic intention
Technology push strategic intention
The company focuses its strategic innovation intentions on a specific customer need. To fulfil this need, the best technology is looked for.
The company focuses its strategic intentions on the development of a specific product for fulfilling customer needs in mainly known markets based on mainly known technologies
The company focuses its strategic intentions on the development of a core technology that is leveraged for satisfying different customer needs in independent domains.
Example
The strategic intention is to develop products that satisfy the customer need of cooking
Example
The strategic intention is to develop microwave ovens
Example
The strategic intention is to develop products based on microwave technology
Cooking
Oven
Microwave oven
...
Heat food
Microwave technology
Infra red
...
...
Cooking
Heating
Microwave oven
...
...
Heater
Heat water
Microwave technology
...
Cooking
Heating
Chemical testing
Microwave oven
Heater
Throughput scanning
...
Rotate dipole
Microwave technology

Figure 5.8 The definition of the functions relates to companies' strategic intentions

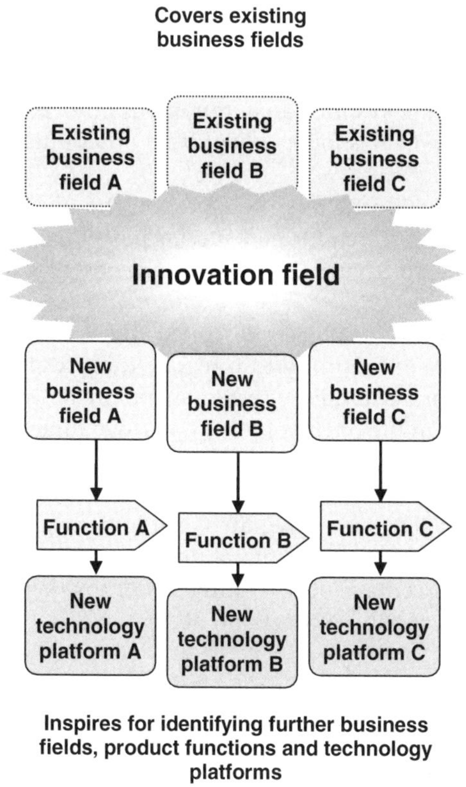

Covers existing business fields

```
┌─────────┐ ┌─────────┐ ┌─────────┐
│Existing │ │Existing │ │Existing │
│business │ │business │ │business │
│field A  │ │field B  │ │field C  │
└─────────┘ └─────────┘ └─────────┘
```

Innovation field

```
┌─────────┐ ┌─────────┐ ┌─────────┐
│  New    │ │  New    │ │  New    │
│business │ │business │ │business │
│field A  │ │field B  │ │field C  │
└─────────┘ └─────────┘ └─────────┘
     ↓           ↓           ↓
 Function A  Function B  Function C
     ↓           ↓           ↓
┌─────────┐ ┌─────────┐ ┌─────────┐
│  New    │ │  New    │ │  New    │
│technology│ │technology│ │technology│
│platform A│ │platform B│ │platform C│
└─────────┘ └─────────┘ └─────────┘
```

Inspires for identifying further business fields, product functions and technology platforms

Figure 5.9 Innovation fields

company intelligence (see Figure 5.9). Lang (1998b) proposes that definition of innovation fields focuses the identifying of potential new successes. Such innovation fields can be understood as broader business fields (Silverstein, 2003: 111). An example is the business field of 'mobile phones', which can be defined as an innovation field of 'mobile communication'. Or 'lacquers and colours' would be changed into the innovation field 'industrial coating'. This broader description of business fields fosters creativity for new related innovations (Silverstein, 2003: 111). Therefore, the purpose of management of innovation fields is to work out customer needs, product functions and technology platforms where at least one of these dimensions is still unknown (Schlegelmilch, 1999).

Specifically, an innovation field should fulfil four criteria, according to (Schlegelmilch, 1999):

- An innovation field comprises the potential to identify at least a new potential business field, product function or technology platform
- An innovation field expresses a competitive advantage
- An innovation field shows an autonomous innovation potential
- An innovation field is independent from other innovation fields.

Such an innovation field is therefore, on the one hand, a starting point for a creative search for new success positions and, on the other hand, according to Savioz (2002), a future observation area for the intelligence.

Additionally to all the objects described above, innovation architecture provides the opportunity to integrate **quantitative key figures**, which helps to understand the system in a more detailed light. Figure 5.3 shows an example of the object knowledge dimension where, on the left-hand side of each object, the number of knowledge carriers is indicated. On the right-hand side, the budgeted costs for developing this object knowledge are integrated. These key figures can be completed by other quantitative information, such as the required development time or percentage of R&D intensity. These key figures depend on the company specific context and needs.

Methodological knowledge

The methodological knowledge dimension visualizes the knowledge needed to generate cognition in scientific research, to develop it into an invention in terms of technical conceptualization, and to introduce it into a market by means of technical and business realization,[1] as shown in Figure 5.10. This figure is a conceptual and exemplarily demonstration of possible levels in the methodological knowledge dimension. It is obvious that some companies will not have a level of scientific research, and other companies will divide the technical conceptualization into the levels of applied research and technology development. Therefore, the number of levels here depends on the number of levels in the object knowledge dimension given by the knowledge depth.

In the methodological knowledge dimension, knowledge depth and the innovation width is visualized. Innovation width is the variance of the knowledge to act and to comply with an action (see Figure 5.10). For example, a company that does research has to have the knowledge about scientific research cascade to search for results in experiments and the knowledge to search in literature for already existing experimentations and results. This company has two different methodological

Concept 83

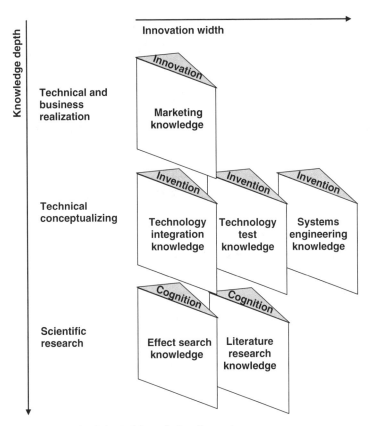

Figure 5.10 Methodological knowledge dimension

knowledge segments on this particular cascade, and this is therefore the innovation width of the company.

A visualization of the coherent combination of object knowledge and methodological knowledge can be seen in Figure 5.11.

Meta-knowledge

The meta-knowledge dimension is the foundation of innovation architecture. It is the knowledge about the object and methodological knowledge. Some authors that see meta-knowledge as the most important value of all knowledge categories (Ward, 1998: 10). According to Ward (Ward, 1998:12ff.), to identify the meta-knowledge in a company, four questions must be answered: According to ward (1998: 12ff)

84 Structured Creativity

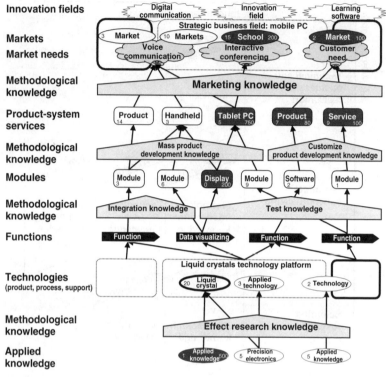

Figure 5.11 Example of an innovation architecture including object and methodological knowledge

- Know who? (e.g. Who around has knowledge about liquid crystal technology?)
- Know how? (e.g. How do we get the ideas for new innovations?)
- Know where? (e.g. What conferences should be attended, in order to stay up to date?)
- Know why? (e.g. What are the cultural values of this place? What is the baggage? What is the vision?).

In each case, effective knowledge is knowing how to go about assembling the relevant object and methodological knowledge to inform a particular decision or judgement (Ward, 1998:13).

To integrate meta-knowledge into an innovation architecture, specific meta-knowledge and general meta-knowledge have to be differentiated. Specific meta-knowledge is always linked to a specific or cluster of objects or methodological knowledge represented in the innovation architecture. This meta-knowledge is a kind of informational specification catalogue that can be directly integrated into the innovation architecture or is shown on a separate document which aims to detail the different objects. In contrast, general meta-knowledge is related to the whole innovation architecture. Therefore, this kind of meta-knowledge does not give additional information for a decision about a specific object of methodological knowledge, but rather is more a representation of how the company as a whole system develops innovations.

Both of the meta-knowledge groups consist of five aspects:

- The source of knowledge (e.g. The liquid crystal technology knowledge comes from partner XY and Dr. Muller is our internal expert)
- The reliance of knowledge (e.g. The journal XY is highly reliable in describing new insights in the domain of nano-technology)
- The importance of knowledge (e.g. Knowledge about liquid crystal technology, installed in tablet PCs, can provide a high return in five years)
- The evolution of knowledge (e.g. Liquid crystal technology knowledge will be substituted by OLED technology)
- The cognitive capabilities to develop new knowledge (e.g. Our internal development department is not capable of developing a tablet PC, but our competitors are).

Conclusion

Innovation architecture visualizes the complexity and systemic interaction of the innovation system because of the linked integration of the innovation-relevant knowledge (object, methodological and object knowledge). Additionally, the evolution of the innovation system can be understood by integrating the planned innovation-relevant knowledge, defining the appropriated functions and innovation fields, thereby allowing the identification of new potential activity fields.

As summarized in Figure 5.12, innovation architecture is a basis for taking decisions based on the visualized strengths (existing knowledge) and weaknesses (knowledge gap) in the innovation system. Through defining the business fields, technology platforms, innovation fields and functions, a strategic direction is given and therefore it is a basic instrument for strategic steering. Innovation architecture also forces the innovation management personnel to think in detail, on every cascade, about the influence of a potential innovation opportunity. It therefore helps to analyze the detailed extension of an innovation

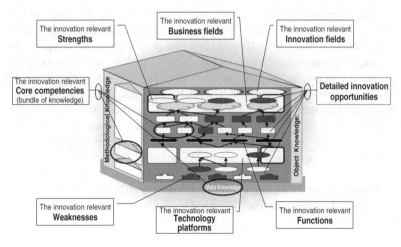

Figure 5.12 Major results of the innovation architecture

opportunity. Last, but not least, innovation architecture is a kind of a strategic architecture[2] which shows the core competencies of a company. Thereby competencies are understood as a bundle of knowledge – object, methodological and meta-knowledge. Often these (core) competencies describe capabilities that result from organizational

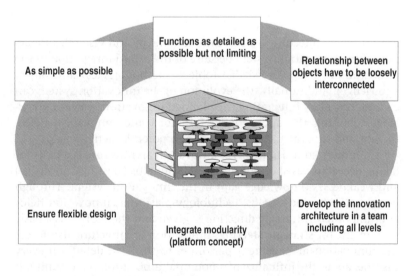

Figure 5.13 Design guidelines for developing an innovation architecture

learning over time and can provide competitive advantage and generate rents (Teece et al.,1997).

In spite of these multifaceted advantages of innovation architecture, it has to be mentioned that the architecture – as every tool in management – has to be up-dated regularly. Without this, the innovation architecture will generate no appreciable advantage. To ensure that innovation architecture is advantageous, it is important to consider some basic guidelines (see also Figure 5.13):

- Innovation architecture has to be designed as simply as possible. It should therefore not be the aim to integrate all the details of the innovation system, but to integrate only the relevant elements for taking strategic decisions.
- The functions should be defined in detail without limits. If the descriptions of the functions are too limited, the focus of development activities is too limited. If the functions descriptions are not detailed enough, they cannot be used for strategic steering or as an identification tool because of too high a level of abstraction.
- The relationships between the objects have to be loose to ensure the autonomy of an object in taking a decision. When the objects are not loosely interconnected, a decision cannot be made about a single object.
- Innovation architecture has to be designed to be flexibly. This means, on the one hand, that the tool with which the innovation architecture is designed (e.g. PowerPoint) should allow changes to be integrated in a short time; on the other hand, the innovation architecture itself should be designed by thinking ahead to integrate changes.
- Innovation architecture should integrate the principles of modularity on each cascade to give R&D the opportunity to leverage an object for another purpose.
- Innovation architecture is a tool for detailed visualization, considering innovation opportunities from a holistic point of view. Therefore, it is important in the process of architecting to integrate employees from marketing and from development, as well as from research.

In summary, innovation architecture is a powerful tool for understanding the innovation system and for preparing strategic decisions. Such strategic decisions are mainly taken in the innovation strategy formulation process. This is the subject of the next section.

Innovation strategy formulation process

In this section, an answer to the second and third research questions regarding designing and implementing a structured innovation strategy formulation process, including the innovation architecture based on the second and third working hypotheses (see chapter 4.2) can be found.

Introduction

Before developing an innovation strategy formulation process, the process has to be integrated into the company's value defining processes.[3] This gives a clear picture of the interfaces with other processes and fosters the development a holistic and integrated strategy formulation process. Such holistic integration into the value defining processes is presented in Figure 5.14.

On a normative level, primary decisions must be made according to the long-term goals of the enterprise. This requires the development of a consistent company policy' and the derivation of an innovation policy. At the same time an awareness of the culture permeating the company is essential. Company culture includes the values held collectively by its employees, which is expressed, for example, in how employees identify with company goals and in the company's behavior towards the environment, and manifest themselves in the company's ability to change and innovate. On the normative level it is not only the making of long-term decisions which is vital for the company's future. Just as essential is who makes these decisions. This question involves the upper level decision-making structures of the company. The guiding principle for the normative level is the principle of meaningfulness. Criteria for meaningfulness refer to the potential of products and services to provide substantial contributions to societal and individual values such as organizational viability, quality of life and personality development.

On the strategic level it is essential that company and innovation policy be transposed into comprehensible strategies. Therefore, based on the strategic intelligence (see p. 91), which formulates on the one hand the company's requirements for information and, on the other hand, analyzes this information, the innovation strategy formulation process is triggered. This process starts with the 'identify' phase (see p. 92), which revises the innovation portfolio for to the purpose of understanding the actual situation, identifies new opportunity fields[4] and details these opportunity fields in a cyclic procedure. In the 'evalu-

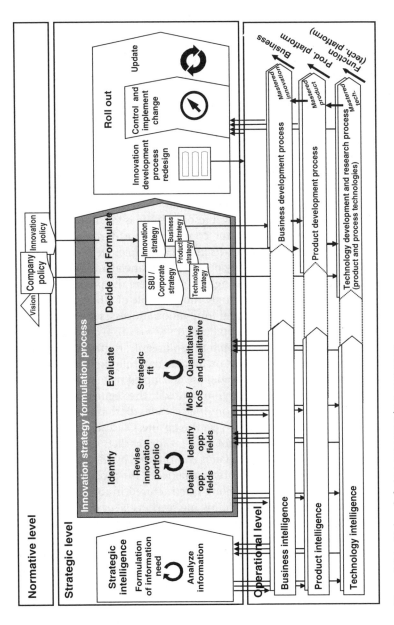

Figure 5.14 Value defining processes in a company

ate' phase (see p. 100) the innovation opportunities are evaluated by analyzing the strategic fit, assessing qualitative and quantitative key figures, and setting the direction for 'make or buy/keep or sell'. This evaluation is the basis for deciding and formulating the innovation strategy in the next phase (see p. 118). With this innovation strategy, which should be based on the corporate strategy, the innovation strategy formulation process as such is closed. Nevertheless, it is still a strategic task to roll out this innovation strategy in a further phase (see p. 121). In this 'roll out' phase, it is firstly essential to redesign the operational innovation development processes to the new strategic requirements in terms of 'structure follows strategy' (Chandler, 1962). Secondly, the implementation has to be controlled and changed if necessary, and thirdly, the innovation strategy has to be updated if major external or internal changes occur.

Finally, on the operational level responsibility is taken for transforming strategies into practice in the context of short-term goals. Operational management expresses itself, for example, in concrete R&D projects in which the necessary personnel, financial and instrumental resources are deployed according to a plan. Here the pointer is 'doing things right', implying accordingly the principle of efficiency. On this operational level, the individual innovation processes are arranged in cascades and segments based on Schaad's (2001: 104) concept. Therefore, an innovation process represents an organizational responsibility domain. The cascades are based on an aimed decoupling of the field of activity. In Figure 5.14 the cascades are decoupled by business, product and technology field of activity. In contrast, the segmentation of the processes is necessary because on one cascade, there is a need for two separate processes. For example, on the business cascade it is essential to have a business development process for markets A and B. Additionally, each process is steered and coordinated by the strategic level because of the asynchronous interface between the processes.[5]

Because the subject of this book is 'structured innovation strategy formulation', the focus in the following sections is set on the strategic level of the value defining processes, describing in detail the three phases of the innovation strategy formulation process shown in Figure 5.14. To ensure the integrity of the description of the strategy formulation process, the 'strategic intelligence' phase and the 'roll out' phase are described in separate sections but on a more general level.

For a more practitioner-oriented understanding of the strategy formulation process, a virtual company which offers digital cameras, named Pixel AG, is used. This allows a demonstration, from a practical

Company Description	PIXEL AG
Revenue in 2003:	2 bn EUR; EBIT: 1.8 mn EUR; # employees: 10.500;
Headquarters:	Zurich, Switzerland, Zurichbergstr. 18
Business:	Digital cameras
Vision:	*CAPTURE AND SHARE YOUR SEEN MOMENTS*
Mission statement:	Based on our current core competencies in the fields of digital photography, we aim to become a full solution provider to our customers. Thus, in line with our vision, we aim to offer the full range of functions to enable the capturing and sharing of your seen moments.
Corporate strategy:	Our strategic goal is to become number two in the worldwide market for amateur digital cameras, with a revenues of 2.5 bn EUR in 2007 and an EBIT of 10%. These objectives will be realized by improving our profitable core, by extending our current activities, by embracing technology push innovation and by pursuing inorganic growth opportunities.

Figure 5.15 Description of the virtual company Pixel AG

point of view, of how to proceed throughout the innovation strategy formulation process. The specific Pixel AG examples during the following sections are always characterized by a box containing the Pixel AG logo on top right. Figure 5.15 presents the description of Pixel AG.

Strategic intelligence

The strategic 'intelligence' step on the strategic level is an ongoing task that guides strategic management in the collection, analysis and application of information that describes relevant facts and trends (opportunities and threats to the organization's entire environment) which is then used to support the strategy formulation process (cf. Savioz, 2002: 33). It is not the task of strategic intelligence to discover these facts and trends by using intelligence tools, such as the publication/patent frequency analysis, the S-Curve analyzes or delphi studies[6] which is more the task of operational intelligence. Rather it is the task of strategic intelligence to formulate the information requirements and to analyze the sampled information holistically.

A tool that supports this management task is the so-called opportunity landscape (Savioz, 2002: 123ff.; Savioz, 2003: 193ff.), which is used to manage the information fields requirements on a competencies basis. The opportunity landscape is based on the gatekeeper approach (Allen, 1977) and constitutes an organizational knowledge base of facts and trends in a company's environment. Measures for the management of these competencies can then be derived from this knowledge. The concept of the opportunity landscape is presented step by step based on Savioz's (2003) descriptions.

First, an inventory of the present knowledge domains has to be created. This should be complemented with domains that could be relevant in the future. The determination of the current and additional domains can be performed in two ways: top-down and bottom-up. In the top-down approach, one derives as far as possible, the strategic knowledge areas and the constituent knowledge domains from the company strategy that are typically formulated by top management. In the bottom-up approach, employees from different departments (research and development, marketing, production, and so on) and from different management levels are brought together in workshops where, via brainstorming, they determine, consolidate and approve possible relevant knowledge domains. The knowledge domains are then broken down into knowledge fields.

The knowledge domains correspond to the competencies that should be built up in the future and the ones that should be maintained and further developed. Because at any given time not all knowledge domains are of the same level of importance, observation depths should be determined. That implies that knowledge domains whose relevance is already quite high should be observed more intently than the ones whose importance is only presumed. The opportunity landscape foresees three observation depths, with a decreasing degree of intensity: the game field, the substitute bank and the offspring.

Visualization gives the opportunity landscape a face. This promotes transparency, and hence communication. A good visualization has three characteristics: completeness, simplicity and sustainability. A possible visualization of the opportunity landscape is shown in Figure 5.16, where the domains are introduced as 'issues'.

The opportunity landscape is an organizational knowledge base of facts and trends in the company's environment. It is an alarm system for discontinuities and it is a pro-active idea generator. Therefore, the opportunity landscape is a basic strategic intelligence tool that summarizes trends, internal ideas, customer needs, strategic requirements and competitor activities in the form of specific issues. In the case of Pixel AG, such issues can be, for example, whether a 7 mio. pixel camera or a 12 mio pixel camera should be developed. Concluding, the opportunity landscape is an ideal starting point for the 'identify' step in the innovation strategy formulation process.

Identify

The 'identify' phase aims at identifying new opportunity fields.[7] To make it possible to identify these fields, it is first of all necessary in the

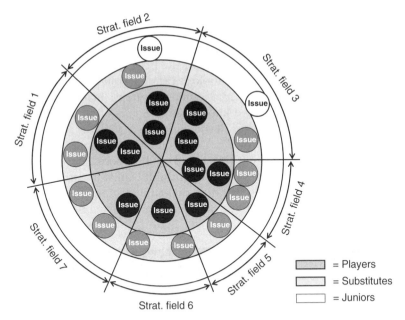

Figure 5.16 Visualization of the opportunity landscape

'revise innovation portfolio' step to understand the information from strategic intelligence. This is based, for example, on different trends, ideas, strategic requirements, customer needs and is summarized as issues in the opportunity landscape. These issues have to be integrated into the innovation architecture (see p. 67). Afterwards, in the 'identify opportunity fields' step, new potential innovation opportunity fields are searched by using creativity methods (see p. 74). To close the identification, the opportunity fields have to be detailed to understand their impact on the innovation system of the company. Therefore, the result is, based on more or less unstructured facts and trends, to build up a structured innovation architecture consisting of all potential innovation opportunities. In the following, the three states in the 'identify' step are described in detail.

Revise innovation portfolio

Based on the opportunity landscape, which represents the results of strategic intelligence and the corporate/business unit strategy, a first draft of the innovation architecture is to be built up in order to understand the actual innovation system and to revise the innovation portfolio. Therefore, the issues of the opportunity landscape, which represent

94 Structured Creativity

ideas of concrete knowledge to be developed, have to be structured into the cascades of the innovation architecture. For example, if there is an issue about the development of tablet PC knowledge, this issue is integrated into the product cascade. Or if the issue is about the development of liquid crystals, the technology cascade would be appropriate. All the issues are clustered into different groups: market, product, module, technology, applied knowledge and scientific knowledge. It should be mentioned that not all companies have activities in scientific or applied knowledge, or that not all products have modules. In this event, these cascades will not appear and can be ignored.

After clustering the issues of the opportunity landscape, they can be linked in order to provide a first draft of the innovation architecture. It will appear from this that some issues cannot be linked because some

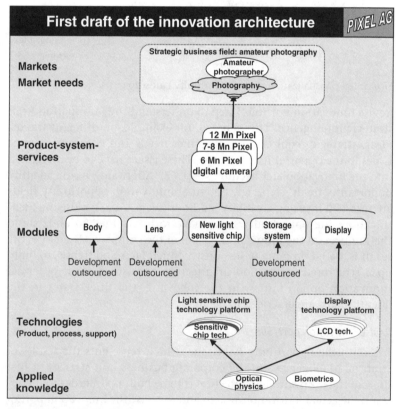

Figure 5.17   First draft of an innovation architecture at Pixel AG

issues are missing in the opportunity landscape. This is for of two reasons: firstly, the objects are integrated in the opportunity landscape in term of issues that need more effort in order to develop them, therefore issues that have been developed in the past are not present. Secondly, the opportunity landscape is a sampling instrument of the results from intelligence. It is therefore possible that objects, obviously missing in the innovation architecture, were not found to be important in strategic intelligence. These missing objects must also be integrated into the first draft of the innovation architecture. In the case of Pixel AG, such a draft is presented in Figure 5.17.

Based on the first draft of the innovation architecture, the functions, innovation fields, business fields and technology platforms must be defined.[8] The definition of these elements[9] is a very important step in innovation strategy formulation. This is because the functions and the innovation fields define the direction of further investigation for steering the innovation system and for identifying new technology platforms and business fields. In the event that the function is over or insufficiently detailed, or the function does not exactly represent the company, an incorrect strategic direction would be defined. The definition of business fields and technology platforms is crucial for clustering the detailed objects, which will afterwards be the categorization of defining innovation orders. Therefore, the definition of these elements should be done with the consensus of all the innovation strategy's decision makers.

Figure 5.18 shows how, in the case of Pixel AG, the identified innovation field and functions can be integrated into the innovation architecture. The innovation field is, in this particular case, the same as the vision; and the functions are mainly the functions of a digital camera.

The following is a checklist for the 'revise innovation portfolio' step:

- Is the corporate and strategic business unit (SBU) strategy integrated?
- Are the customer needs analyzed in detail?
- Are all the innovation relevant trends for products, markets, businesses, technologies known?
- Are the competitors' new products and technologies analyzed and is their future strategic position known?
- Are the gatekeepers involved for identification of internal ideas?
- Are the functions, innovation fields, business fields and technology platforms defined, appropriate to the company needs?

96 *Structured Creativity*

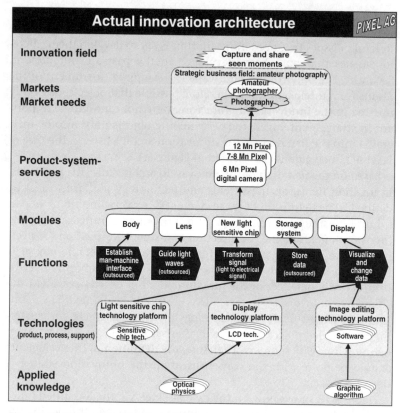

Figure 5.18 Actual innovation architecture at Pixel AG

If all the answers to these questions are positive, new opportunity fields can be systematically identified.

Identify opportunity fields

In contrast to the last step, which was about sampling and defining, this step is creative. Based on the innovation fields and functions, new business fields and technology platforms are identified. In doing so, it is not intended to identify a specific technology or a specific product, because this is a task for strategic intelligence. It is more an aim to identify potential fields of activity which will provide the opportunity for developing specific innovations in the future.

For this purpose, a workshop regrouping the gatekeepers of the company's innovation system is a suitable basis for finding opportunity fields. In this workshop, different creativity methods can be used, such as brainstorming, mind mapping, discussion 66, method 635.[10] When using these methods, the quality is not important; the quantity of the ideas of the workshop members is however, essential to enforce creativity.[11] The workshop should consist of two parts:

- Innovation fields analysis:[12] Based on the preliminary definition of innovation fields, primarily new business fields should be identified. For example, in the case of Pixel AG the innovation field 'capture and share seen moments' allowed the full new product solution to be identified, which means nothing other than offering a digital camera with a service to upload the pictures onto a server to be managed in terms of changing, administrating, sending, and so on..
- Functional analysis:[13] Based on the already defined functions of the innovation architecture, questions have to be answered as to what technology platform also fulfils a specific function and what business field requires this function. For Pixel AG, the function 'visualize and change data' made it possible to identify the technology platform of beaming the data on a flat surface.

The identified opportunity fields can be roughly evaluated at the end of the workshop. It is not the aim to make a quantitative evaluation, but more a qualitative evaluation which only considers if there is a realistic potential to meet a certain opportunity field. The potential opportunity fields should afterwards be integrated into the innovation architecture.

The checklist for the 'identify opportunity fields' step is:

- Are all the gatekeepers involved in this identification workshop?
- Do the gatekeepers have enough information about the preliminary defined functions and innovation fields in order to be able to identify new opportunity fields?
- Are the identified opportunity fields integrated into the innovation architecture?

Up to this point, new opportunity fields are integrated into the innovation architecture, but they have to be detailed in order to evaluate them in a further step.

Detail opportunity fields

It is obvious that as yet the innovation architecture is incomplete, and the level of detail is not the same for all objects. This step therefore aims to detail all of the objects in the innovation architecture to one and the same level. This can be done by considering three aspects.

Firstly, the identified opportunity fields themselves have to be detailed by finding the markets and customer needs for the business fields, and the products and process technologies for the technology platforms. And these specific objects have to be linked to already existing objects or, if this is not possible, the missing objects have to be identified and integrated.

For Pixel AG (see Figure 5.19) it is necessary to develop a specific mini beamer technology and scientific knowledge in 'dichroic physics' is required. Also, in order to develop the full solution of getting an additional homepage module for managing the pictures, a 'Wireless Lan' module for sending the pictures to the server is required. These modules also require specific technologies. Furthermore, this full solution product can be divided into a standard and flexible version. The standard would be for the hobby photographer and the flexible version would be for the professional photographer, which represents a new business field.

Secondly, the innovation architecture has to be completed in terms of its object knowledge. This means that every object has to be evaluated by the level of knowledge that already exists in-house. For Pixel AG (see Figure 5.19) a twofold differentiation is made about object knowledge: (1) The objects where object knowledge exists in the company (white objects) are presented, representing the company's strengths, and (2) objects where this knowledge is not available in the company is also presented (grey objects), representing the company's weaknesses. This twofold differentiation can be expanded by a more detailed categorization. For example a differentiation can also be made for the particular object knowledge that exists in the company. The number of levels of the object knowledge in the innovation architecture has to be adapted to the needs of the company.

Thirdly, the innovation architecture has to be completed with methodological knowledge and meta-knowledge. The methodological knowledge can be integrated as can be seen in Figure 5.19. The meta-knowledge can only be specifically integrated into the innovation architecture visualization as, for example, the quantitative key figures of every object. Other meta-knowledge, such as the source of the knowledge, has to be predented in additional documents. Thus it is not the aim to identify all possible meta-knowledge, but rather to sample the meta-knowledge that is relevant for the company specific needs and for the evaluation phase.

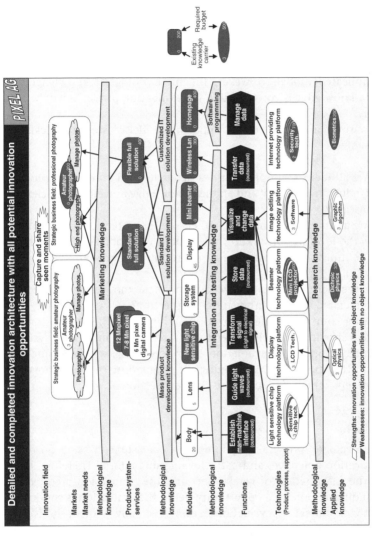

Figure 5.19 Structured and complete innovation architecture with all innovation opportunities at Pixel AG

This final innovation architecture, presented in Figure 5.19, of the phase 'identify' is completely constructed in terms of the described concept of innovation architecture. Because in the next phase (the evaluation phase) the emphasis is on the object, the methodological knowledge and the meta-knowledge are only secondary considerations because they are often a constraint of the objects to be developed. In the following it was decided to use only the object knowledge dimension of the innovation architecture in order to present a less complicated model. However, the methodological knowledge and meta-knowledge is considered in the evaluation phase, as will be seen later.

To summarize the 'detail opportunity fields' step a checklist is presented:

- Are all identified opportunities fields detailed on market, product, module, technology and applied knowledge as well as the scientific cascade?
- Are all the objects integrated to ensure the systemic interaction?
- Is the level of object knowledge about each object integrated into the innovation architecture and are the strengths and weaknesses visible?
- Is the methodological knowledge and meta-knowledge analyzed and integrated into the innovation architecture?
- After completing the innovation architecture, the innovation system should be clearly understood in terms of complexity, systemic interaction and its evaluation. Therefore, this is a highly structured starting point for conducting an evaluation of the innovation system, which is described in the next section.

Evaluate

The 'evaluate' phase aims to prepare a consistent decision on future innovation objectives and paths for focusing efforts. Therefore, based on the information of gatekeepers and the innovation architecture, including all potential opportunity fields, the evaluation is conducted. Firstly, a 'strategic fit evaluation' is done, secondly, a 'quan-titative and qualitative evaluation' is done; and lastly, a 'make or buy/keep or sell evaluation' is done. These three steps should be seen as a circular procedure aiming to prepare an innovation architecture that can be seen as a proposal for a strategic decision.

Strategic fit evaluation

The strategic fit evaluation aims to prove the consistency of the innovation opportunities itself, with its environment, consisting of the value providing system of the company, and the whole environment of the company. To evaluate the consistency, there are three different types of strategic fits, according to Porter (1996: 70f).

- First-order fit is simple consistency between each activity. Which means, in the context of innovation strategy formulation, consistency between the innovation opportunities in the innovation architecture.
- Second-order fit occurs when activities are reinforcing, therefore the innovation opportunities in the innovation architecture fit with the overall activities of the company.
- Third-order fit goes beyond activity reinforcement to optimization of effort. This would be, in the context of innovation strategy, the fit of the innovation opportunities with the future development of the company's environment.

Thus, 'the more a company's positioning rests on activity systems with second- and third-order fit, the more sustainable its advantage will be. Such systems, by their very nature, are usually difficult to untangle from outside the company and therefore hard to imitate' (Porter, 1996: 70f).

Based on these three types of strategic fit, possible tools are presented for each type. These tools are based on existing tools that are shown in alignment with the innovation architecture. The following are not claimed to be a complete tool set for each company, but rather a basic tool set that has to be completed to the company specific needs.

Tools for first-order fit: For ensuring a first-order fit, the innovation architecture itself is a tool to identify the consistency between the innovation opportunities. Basically, this consistency is already ensured in the step 'detail innovation opportunities' by ensuring that all the innovation opportunities are systemic interacting, and therefore that the objects are linked together over the cascades.

Additionally the **functional handshake** can be used to ensure the first-order fit (cf. Biedermann *et al.*, 1998: 549ff.). This tool aims to align the market pull and technology push based activities and ensures their coordination which is, according to several authors (see exemplary Leenders and Wierenga, 2002; Souder, 2004), a major challenge to master. More concretely, the functional handshake is based on the alignment of the customer needs and the product technologies over the product functions. This is optimally achieved with members of marketing and R&D by answering the question as to whether the functions of the product technologies can be used for a customer need or whether functions demanded by customers can be fulfilled by product technologies in R&D. This functional handshake can be done in the innovation architecture, by linking the technology side with the market side by functions as illustrated in Figure 5.20.

Tools for second-order fit: To ensure the second-order fit, and therefore the fit with the company as a whole, the **key success factor**

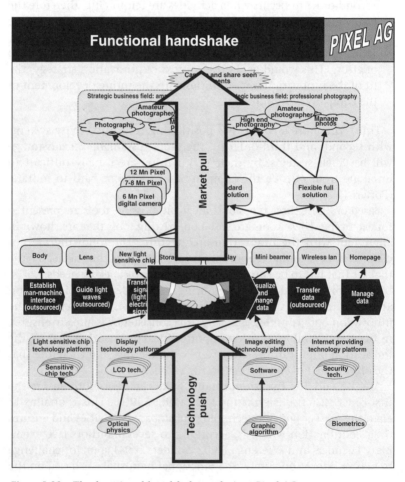

Figure 5.20 The functional handshake analysis at Pixel AG

analysis can be made. This tool's objective is to analyze the company's success factors. These success factors consist of the key consuming factors (which are the factors affecting the end customer in his buying decision), the key buying factors (which are the factors that are influencing the retail), and the key factors for success (which are the factors that differentiate the company from their competitors). For example, in the case of Pixel AG (see Figure 5.21) the key consuming factors for professional photographers are that the homepage has to be adapted to the professional photographers' system, and that it has to allow a professional public appearance. The key buying factors demanded by the camera shops are, eventually, IT support in the case that the homepage has to be adapted. Key factors for success are micro marketing and the innovation rate.

These three categories of key success factor have then to be broken down into competencies that the company has or should have. For example, based on the key consuming factor 'Homepage has to be adapted on own system', the company should have the competence in

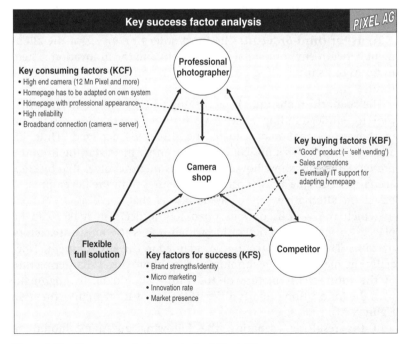

Figure 5.21 Key success factor analysis at Pixel AG

R&D to develop a flexible Homage generator and should supply the camera shop with IT support. Therefore, these necessary competencies express a need for knowledge in the innovation system which should be the same as the object knowledge in the innovation architecture. In the event that this is not so, the innovation architecture has to be changed.

A second tool for enforcing the second-order strategic fit is the **core competence analysis** of Prahalad and Hamel (1990). Because core competencies are a major base of competitive advantage, they have to be enforced by the innovation system. Therefore, innovation opportunities that enforce an existing core competence, that will also be a potential core competence in the future, are often more valuable for the specific company than other innovation opportunities. Thus, the object knowledge of the innovation architecture should be in alignment with the core competencies of the whole company (see Figure 5.22).

At this point, it must be mentioned that it is crucial that the methodological knowledge and the meta-knowledge should be oriented in alignment to the competencies, especially the core competencies. If this is done, the innovation opportunities are fitted to the company in relation to their content, and the second-order fit is ensured.

Tools for third-order fit: The third-order fit consists of the alignment between environmental development and the innovation system in terms of its innovation opportunities to optimize the effort for the future.

The **scenario technique** is, according to Holt (1988: 139f), a first tool for understanding alternative futures, based on a discussion of the events that may lead to the situations depicted. Thus, an attempt is made to set up sequences of events which, starting from the present situation, show how future states might evolve, step by step. Scenario technique aims at providing pictures of future developments based on alternative sets of assumptions that serve as a context, in which the various options open to the decision maker can be placed, so that he can determine which one is the most satisfactory overall.[14] This scenario technique, adapted to the environment, especially the markets and technologies, should provide a sharper picture of the future. This picture of the future should be in alignment with the innovation opportunities of the innovation architecture (see Figure 5.23).

In the resulting scenarios, the following questions should be answered:

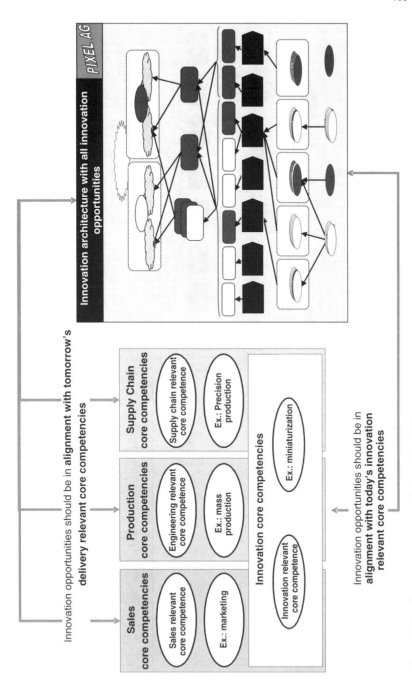

Figure 5.22 Core competence analysis at Pixel AG

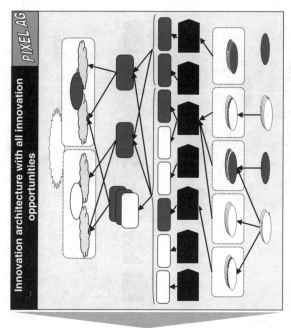
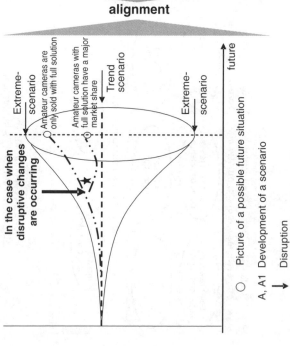

Figure 5.23 Scenario technique at Pixel AG

- Who will the customer be in the next 5 to 10 years? Which customer needs are to be satisfied? How will the customer's needs change in the next 5 to10 years? Which products are concerned?
- How is the market structured? How do the segments differ? What are the geographical differences?
- Which market trends are identifiable?
- How is the role of the suppliers defined? Could a supplier make a forward integration?
- How will technological development change the business in the next 5 to 10 years? What changes in production processes are expected? How will information and data flow change?
- Who will the competitor be? How have they changed in the last 5 to 10 years? Which strategic directions do the competitors have? What are the entrance barriers for the market? What is the importance of national barriers?
- What other restrictions are there? Are future competencies, ecological or labour based restriction visible?

Looking at these questions, it is clear that the scenario technique is a tool in this context which constructs Porter's (1980: 26) five forces model. It is essential for this to be understood in the company context in order to develop innovations and gain future competitive advantage.

A second tool for ensuring the third-order fit is the **market portfolio analysis**. The aim is mainly to understand into which market it is wise to invest. There are several different market portfolio concepts: Two concepts (the 'McKinsey Portfolio Matrix' and the 'Boston Consulting Group Product Matrix') seem to be the most used. The 'McKinsey Portfolio Matrix' 'is an aid for determining the relative position of product lines and diversification projects based on an analysis of their competitive strength and the attractiveness of their markets', therefore it is 'to provide an analytical basis for strategic decisions concerning resource allocation (including possible elimination) for existing product lines as well as for decisions concerning diversification into new business areas' (Holt, 1988: 247). In contrast, the 'Boston Consulting Group Product Matrix' (Henderson, 2003: 42) is an 'aid for determining the relative position of product lines based on their relative market share and the growth patterns of their markets' and therefore it is 'to provide an analytical basis for strategic decisions concerning support or elimination of product lines' (Holt, 1988: 250). The results of the market portfolio analyzed – from the BCG as well as from the McKinsey Portfolio Matrix – should be in complete alignment with the innova-

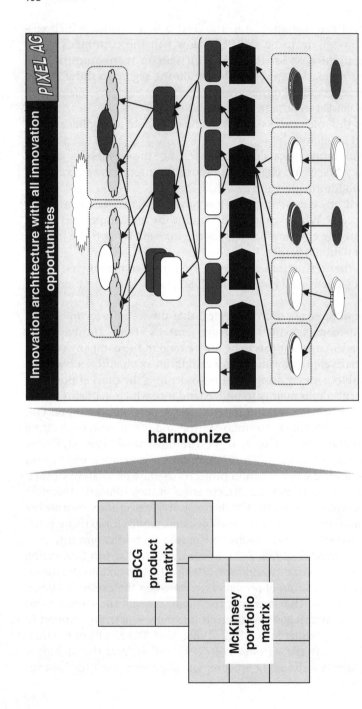

Figure 5.24 Market portfolio analysis at Pixel AG

tion opportunities presented in the innovation architecture (see Figure 5.24).

To summarize the three strategic fits, as a check-list, the following questions should be answered positively:

- Is the consistency between each innovation opportunity and the overall consistency ensured, especially the fit between market pull and technology push?
- Do the innovation opportunities reinforce each other and the company activities in terms of its competencies, especially its core competencies?
- Do the innovation opportunities optimize the efforts for the future, in terms of scenarios?
- Do the innovation opportunities conform to the corporate/business unit strategy?

At this point of the evaluation, the innovation opportunities are analyzed in terms of their fit. Additionally, a quantitative and qualitative evaluation has to be done.

Quantitative and qualitative evaluation

The quantitative and qualitative evaluation aims to find out whether innovation opportunities are feasible in terms of knowledge and time, and whether they are profitable. For this purpose, three tools are presented that answer needs of feasibility and profitability. Additionally, a summarizing tool, the dynamic technology portfolio, is presented.

To ensure the feasibility in terms of knowledge, the **knowledge gap analysis** is a possible tool. The results of knowledge gap analysis indicate to the management where they need to seek new insights, and direct their time and energy (cf. Krogh *et al.*, 2001: 431). But before directly investing in an existing knowledge gap, it has to be determined whether the knowledge could be developed in-house or if it exists externally, and whether the knowledge is so strategically important that it has to be developed in-house. Combining the answers leads to a direction of how every knowledge gap should be managed. In combination with the innovation architecture, this knowledge gap analysis should be undertaken for each (or at least for a cluster of them) object, methodological and meta-knowledge element in the innovation architecture in order to find out the required direction to develop the knowledge for a specific innovation.

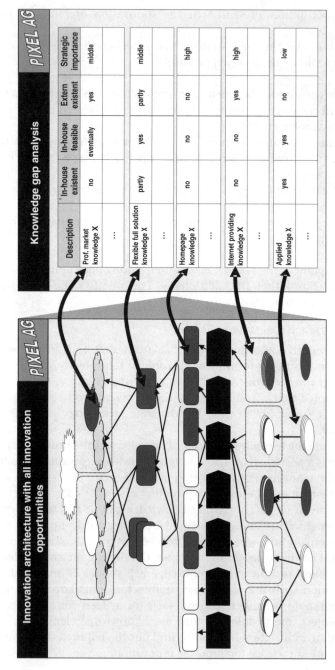

Figure 5.25 Knowledge gap analysis at Pixel AG

Pixel AG, for example, has no knowledge about internet technology. At the same time, it is not possible to develop this technology in-house, although the technology does exist externally. However, it is very important that the development of this technology is done in-house, to gain a sustainable competitive advantage (see Figure 5.25). To ensure the feasibility in terms of time, **the innovation roadmap**,[15] which is the combination of a technology, product and market roadmap, can be used. An innovation roadmap is therefore a tool that comprehensively depicts the assumed development of essential innovation opportunities over time. A roadmap helps understanding of the future and aids synchronization of R&D and marketing in terms of time. Based on the innovation architecture such a roadmap can easily be derived by bringing the object knowledge, which is not completely developed, on a time axis as shown in Figure 5.26. This innovation roadmap shows exactly how long R&D has to develop a technology or a product because the direct relation to the market introduction is visible. Additionally, the innovation architecture helps to determine the influences if something cannot be developed in time. Another advantage of the innovation roadmap is the opportunity to fix the market introduction time, which is given based on customer needs, and then reversed, to see how long a certain project might take from start to finish.

To ensure the profitability of the innovation opportunities a **resource allocation** has to be made, followed by a profitability analysis. The resource allocation can be made on the basis of the innovation architecture by allocating to a bundle of object, methodological and meta-knowledge resources. These resources include human resources as well as financial resources. To ensure that the definition of the number of resources is maintained, a person must be designated for each resource bundle. Such a resource allocation can be visualized as shown in Figure 5.27.

In addition to this resource allocation, the return on investment has to be analyzed. Allocated resources are accounted for, so that a specific customer need can be reached. It should be mentioned that the costs of developing technologies or products that are used in several markets have to be divided into the different markets. Also, the possible market income is analyzed over time. With this data, an investment or payback calculation can be made, which is quite a common method, or the more recent method of the net present value calculation NPV, can be used on the basis of the discounted free cash flow analysis, according to Rappaport (1986). The NPV 'represents numerical values refer-

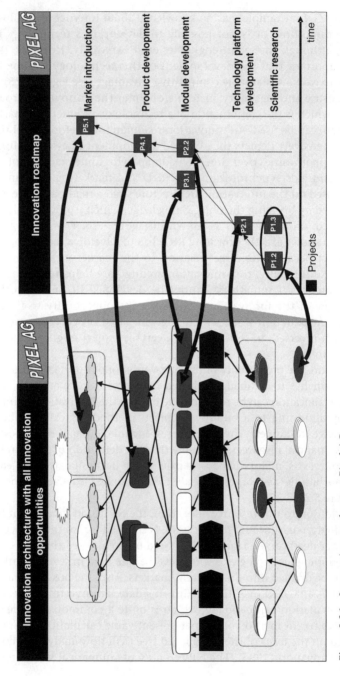

Figure 5.26 Innovation roadmap at Pixel AG

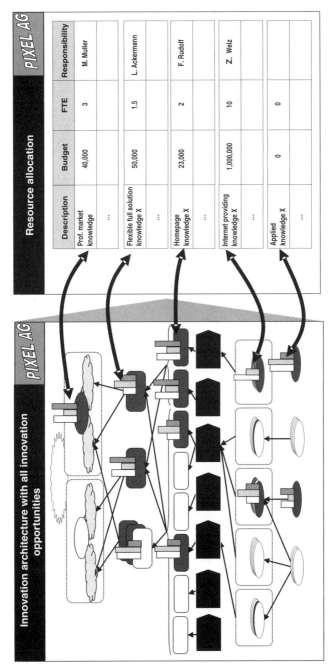

Figure 5.27 Resource allocation at Pixel AG

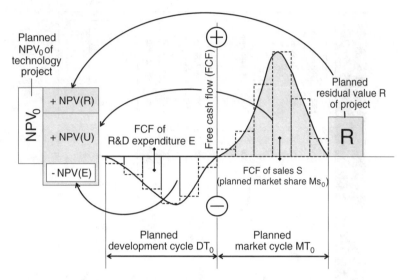

Source: Tschirky (1998a: 348)

Figure 5.28 Establishing R&D projects' net present values

ring to increases or decreases of the total company value. It is evident that, through this procedure, the interest of top management' in innovation strategies 'and R&D projects is much higher than in financial project data which only expresses a local view from the R&D department'. The procedure by which to develop the NPV is shown in Figure 5.28.

In addition to these calculation methods, several other key figures can also be calculated.[16] The key figures that are actually used depend on the company specific context.

To conclude, the three tools – knowledge gap analysis, innovation roadmap and resource allocation with investment calculation – are a good basis for making a qualitative and quantitative evaluation. Nevertheless, these results produce a large amount of information and therefore have to be summarized for presentation to top management. For this purpose the **dynamic technology portfolio** is an appropriated tool (see Figure 5.29). This portfolio rates and positions all major technologies according to their 'technology attractiveness' with respect to their innovation and market potential, and their corresponding 'technology strength', in example the resources currently available within the company. The data for filling the portfolio should be available from the analysis done in the previous stage of the innovation strategy formulation process.

Once the portfolio has been developed, its strategic evaluation can take place. This focuses on setting priorities as to the promotion or reduction of technology development resources or even the phasing-out of aging technologies. The latter decision usually follows intensive internal discussions. In particular, consensus has to be reached on core technologies. They constitute strategic knowledge assets of companies and are usually developed in-house with high priority. Additionally, the dynamic technology portfolio integrates technologies that are attractive despite the lack of company resources.

The main merit of the technology portfolio lies in its high degree of condensation of strategic information and at the same in its ease in communicating strategic decisions. In addition, a successfully finalized technology portfolio reflects completion of a constructive collaboration between experts from R&D, production and marketing, which is a valuable goal on its own.

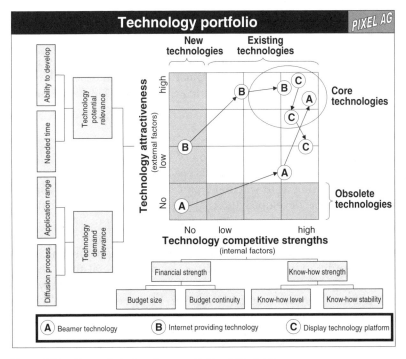

Figure 5.29 Dynamic technology portfolio at Pixel AG

Pixel AG's simplified technology portfolio (see Figure 5.29) shows that the display technology based on LCD is today a core technology, but in the future will lose its attraction. In contrast, the internet technology is today already quite attractive, but this attractiveness will continue to increase in the future, so it would potentially be a core technology.

Concluding the technology portfolio summarizes the results of the 'quantitative and qualitative evaluation' step and gives a proposition for strategic goals for the innovation system. The following questions should summarize the essential questions to be answered in the quantitative and qualitative evaluation:

- Is the knowledge gap and its strategic importance for each innovation opportunity known?
- Is the planning time necessary to develop an innovation opportunity known in detail?
- Are the financial key figures, required in the company context, for the innovation opportunities known?
- Is the dynamic technology portfolio meaningful?
- Can a decision be taken in terms of which way to go?

Together with the evaluation of the strategic goal, which was the objective of this evaluation step, evaluation of the strategic path is also essential for deciding and formulating a strategy.[17] This is the subject of the next section.

Make or buy/keep or sell evaluation

In the previous section the question 'Which way to go?' was central, now the question 'Make or buy/keep or sell' is essential, meaning making a decision as to whether a part or a whole innovation opportunity is to be developed internally or externally, and if existing objects in the innovation architecture should be kept or sold.

The make or buy decision considers the following aspects, according to Brodbeck (1999: 99):[18] limited resources, development time, fixed costs, coordination, sourcing alternatives and cultural fit to the co-operation partners, and so on.. Many of these aspects were already considered in the previous steps, which can be taken as support for the preparation of the make or buy decision. A summary of the reasons for a make or buy is presented in Figure 5.30.

According to Brodbeck (1999: 114),[19] the keep or sell decision considers aspects, such as the internal return on investment, the possibility

Strategic reason for 'Make'	Strategic reason for 'Buy'
• Development of core competencies • Achievement of a technology leadership • Control of the entire innovation process • Independency • Exclusivity of possession • Decision freedom in terms of development paths • Decision freedom in terms of object utilization • Prestigious image advantages to customer • Conserve opportunity of clear technology orientation to the company	• Object with low competitive strategic impact • Aspiring technology presence • Split of technology risk • Reduction of fixed costs • Reduction of time need until the technology is available • Utilization of synergies • Conservation of high flexibility in the organization

Strategic reason for 'Keep'	Strategic reason for 'Sell'
• Object as core competencies • Exclusivity of possession • Generation of many own products on the basis of the technology • Exploitation of monopoly position • Decision freedom in terms of market handling • Gain prestige an image advantages in front of customers	• Object with low competitive strategic impact • Access to new targeted markets • Highest possible amortization of technology investments • Standardization of objects • Generation of high amount of applications of the technology • Gain technology leadership • Protection of own resources

Source: Adapted from Brodbeck (1999: 101, 116)

Figure 5.30 Reasons for make or buy/keep or sell

of selling as a basis for a higher market development, possible utilization alternatives, joint use of R&D learning curve, and so on. Some of the aspects were already treated in previous sections. These results and some additional information are enough to propose a decision in this context.

In the case of Pixel AG such make or buy/keep or sell evaluation leads to the rejection of the market of professional photography, and therefore the flexible full solution, due to the fact that the IT services cannot be developed by Pixel AG. Additionally, it was decided to buy the function 'manage data' because it is important to have it in-house, but at the same time there are hardly any available competencies in-house.

The discussion of make or buy/keep or sell[20] should include all objects, methodological and meta-knowledge clustered in groups. Therefore, at the end of this step the following questions, in terms of a bundle, should be answered:

- Has each innovation opportunity of the innovation architecture been analyzed in detail regarding how to develop it? Make it internally or buy externally?
- Is each bundle of objects, methodological and meta-knowledge analyzed to determine if it is still needed? Keep it internally or should it be externally sold?
- Is the strategic path for each innovation opportunity known?

Decide and formulate

Decide the innovation strategy

In this section, the evaluation phase leads to a proposition for taking a strategic decision containing the strategic goals, in the form of the innovation opportunities that should be developed, and a strategic path, consisting of a 'make or buy/keep or sell' proposition.

Nevertheless, this mainly fact based strategic decision has also to be in alignment with soft factors (Hunger and Wheelen, 2002: 115). Such soft factors can be embedded by integration of the important interest groups of such decisions according. The most important interest groups to consider are presented in the following, based on Hunger and Wheelen (2002: 115):

Interests from Stakeholders: the attractiveness of a strategic alternative is affected by its perceived compatibility with the key stakeholders in a corporation's task environment. Creditors want to be paid on time. Unions exert pressure for comparable wage and employment security. Governments and interest groups demand social responsibility. Stockholders want dividends. Management must consider all of these pressures in selecting the best alternative. To assess the importance of stakeholder concerns in a particular decision, strategic managers should ask four questions: (1) Which stakeholders are most crucial for corporate success? (2) How much of what they want are they likely to get under this alternative? (3) What are they likely to do if they do not get what they want? (4) What is the probability that they will do it? Strategists should choose strategic alternatives that minimize external pressures and maximize stakeholder support. In addition, top management can propose a political strategy aimed at influencing key stakeholders. Some of the most commonly used political strategies arc constituency building, political action committee contributions, advocacy advertising, lobbying and coalition building.

Interests from the corporate culture: if a strategy is incompatible with the corporate culture, it probably will not succeed. Foot-dragging and even sabotage could result, as employees fight to resist a radical change in corporate philosophy. Precedents tend to restrict the kinds of objectives and strategies that management can consider seriously. The 'aura' of the founders of a corporation can linger long past their lifetimes because they have imprinted their values on a corporation's members.

In considering a strategic alternative, strategists must assess the strategy's compatibility with the corporate culture. If there is little fit, management must decide if it should (1) take a chance on ignoring the culture, (2) manage around the culture and change the implementation plan, (3) try to change the culture to fit the strategy, or (4) change the strategy to fit the culture. Further, a decision to proceed with a particular strategy without a commitment to change the culture or manage around the culture (both very tricky and time-consuming) is dangerous. Nevertheless, restricting a corporation to only those strategies that are completely compatible with its culture might eliminate the most profitable alternatives from consideration.

Interests from key managers: Even the most attractive alternative might not be selected if it is contrary to the needs and desires of important top managers. People's egos may be tied to a particular proposal to the extent that they strongly lobby against all other alternatives. Key executives in operating divisions, for example, might be able to influence other people in top management to favour a particular alternative and to ignore objections to it.

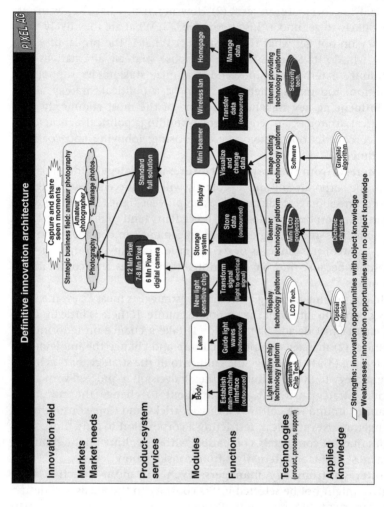

Figure 5.31 Definitive innovation architecture at Pixel AG

People tend to maintain the status quo, which means that decision makers continue with existing goals and plans beyond the point when an objective observer would recommend a change in course. People may ignore negative information about a particular course of action to which they are committed because they want to appear competent and consistent. It may take a crisis or an unlikely event to cause strategic decision makers to consider seriously an alternative they had previously ignored or discounted.

At the point that all important interest groups concerned with innovation management are known and were involved, a decision based on the innovation architecture can be taken.

In the case of Pixel AG, the decision was taken as the strategic proposal targeted. This leads to the definitive innovation architecture as can be seen in Figure 5.31. This decision has then to be retained by formulating the innovation strategy in the form of an official document.

Formulating an innovation strategy

The formulation of the innovation strategy has, firstly, the aim of retaining the essential aspects as to why such a decision has been taken. Secondly, this innovation strategy is the official document for the company, which has to be considered on a strategic as well as on an operational level in taking further decisions. In short, the innovation strategy document is the plan for directing the innovation system to a future intended position. An example of the content of the overall documents of an innovation strategy is presented in Figure 5.32 and Figure 5.33. For each of the six example sections, the content is presented on the left-hand side, and on the right-hand side, an abstract of Pixel AG's innovation strategy. In Figure 5.34 a concrete example of an innovation strategy is presented.

Rollout

Redesign innovation development processes

Based on the changed innovation strategy, the operational innovation development processes have to be redesigned if necessary. The aim is to develop organizational innovations to increase effectiveness and efficiency, in addition to the technological and business innovations. Therefore the processes and structures should be in alignment with the technological and business innovation opportunities. In this context, Tushman *et al.* (1997: 13) said: 'Management teams must be able not only to craft strategic intents, but also to directly couple their strategic intents to organizational architectures.'

| **Initial situation** | Innovation opportunities | Strategy and gaps | Innovation objectives | Innovation strategy | Strategic plan |

Within the initial situation the current vision of the company is described. Furthermore its strengths such as core competencies and weaknesses in the past are shown as well as the environmental, social, technology,… trends are shown.

Information source: strategic intelligence

Vision: 'Capture and share seen moments'
Trends: The increasing demand for digital cameras with more Mn. pixel can only be met by important R&D investment
Strength: Pixel AG's culture excels in developing uncommon ideas into innovations
Weakness: Pixel AG is not in the position to hold for the long term the price/performance for…

| Initial situation | **Innovation opportunities** | Strategy and gaps | Innovation objectives | Innovation strategy | Strategic plan |

Specific and potential fields of innovation opportunities should be presented, to demonstrate and retain the potential varieties of future activities

Information source: identify step

Specific Opportunities
-A 7-8 Mn Pixel camera in the short-term and 12 Mn Pixel camera in the middle-term
-…
Opportunity fields:
-A full solution product that contains a camera as well as the service to manage the picture afterwards.

| Initial situation | Innovation opportunities | **Strategy and gaps** | Innovation objectives | Innovation strategy | Strategic plan |

The gaps should be shown that are needed to develop the innovation opportunities. It is essential to show the gap between the opportunity and its strategy fit as well as the gap to develop it. Furthermore, a consideration of make or buy / keep or sell options should be presented.

Information source: evaluate step

Gaps: A flexible full solution product for professional photographers will require an additional IT after-sales service. Knowledge for developing a homepage for picture management, does not exist in-house.
Buy: The knowledge of managing pictures through a homepage has to be bought because of its high strategic importance.
Make:…

Figure 5.32 Innovation strategy content including a simplified example of Pixel AG

| Initial situation | Innovation opportunities | Strategy and gaps | **Innovation objectives** | Innovation strategy | Strategic plan |

The innovation objectives, decided by the members of the innovation strategy formulation process, should be described in detail. Furthermore, the motivation of the decision should be presented

The standard full solution product will be developed. This is based on the fact that competition in the pure digital camera business is increasingly dramatically. Therefore Pixel AG will differentiate its product with a full solution combining the camera and the homepage based picture management system.
....

Information source: decision protocol

| Initial situation | Innovation opportunities | Strategy and gaps | Innovation objectives | **Innovation strategy** | Strategic plan |

The innovation strategy is the central document which should be communicated. Is should consist of three elements:
1 What are the innovation objectives?
2 How does this innovation objective contribute to the general strategic objectives and what is the benefit for the company (short-,middle-and long-term)?
3 What is the general impact for each company department?

Innovation objective: Standard full solution product
Strategic contribution: The standard full solution product will support the strategic intention to extend activities based on actual products. The revenue will increase by approximate 0.3 mio bn EUR until 2007.
Impact: A company, developing homepages and data management systems will be bought.

| Initial situation | Innovation opportunities | Strategy and gaps | Innovation objectives | Innovation strategy | **Strategic plan** |

To close the gap between innovation strategy and its implementation, a detailed strategic plan should be derived for each innovation objective consisting of:
-Clear and consistent objectives
-Required human and financial resources
-Detailed innovation roadmap and milestones
-Responsibilities

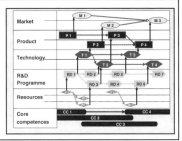

Innovation strategy of Pixel AG

PIXEL AG

Our strategic objective:

Recognizing photographers' increasing need for managing photos, our wish is to be the leader in the domain of providing a means of sharing and managing digital photos. This full solution product allows our customers more easily to share pictures. The aim is to share visual images as simply as possible by reducing the steps necessary to the minimum and increasing the possibilities of sharing to the maximum.

Our strategic path:

This strategic objective will change our understanding in developing new products. In the past the main performance driver for our products was the number of pixels and the overall quality of the picture itself. In the future, the main performance driver will be the number of required steps and the increased possibilities of sharing visual images.
This new understanding has an impact on our innovation activities:

- The technological knowledge of developing management software, as part of our full solution product, has to be acquired. A possible solution to this problem is the acquisition of a company that is active in the domain of managing software and the internet. A second possibility is a strategic collaboration with such a company. The aim is to acquire the relevant knowledge before the end of 2004 and to develop the technology by the end of 2005.
- Since our competitors have successfully developed the 12 Mn Pixel chip, we have definitely lost our third position in the market. Based on our strategic goal of offering a full solution digital camera, the importance of the number of pixels is reduced for our future products. Therefore, the development of technologies for transforming light signals into digital signals will be reduced in importance for the company. Nevertheless, an important investment will still be necessary in order to not completely lose contact with the competition. The aim is no longer to be the first in the high end market but to be present in the middle end market.
- In the middle of 2006, we want to offer a full solution product. This full solution product should contain a data transfer system (eventually Wireless Lan) with a connection to an online data management system with all photos managed directly from the camera.

In its quest to achieve excellence in capturing and sharing visual moments, Pixel AG is moving to build on the investments already made in R&D, to make our strategic goal possible. The realization of this strategy will allow our future basis to be built up for competitive advantage.

Figure 5.33 The innovation strategy of Pixel AG

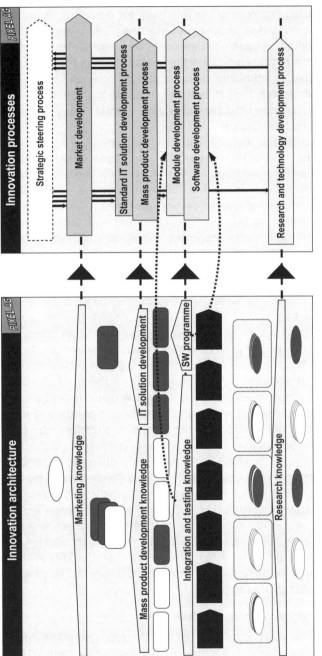

Figure 5.34 Deriving the innovation processes, based on the innovation architecture at Pixel AG

The underlying concept in this book for designing an innovation organization is based on Schaad's (cf. Schaad, 2001) innovation process understanding. Therefore the processes are essentially designed on that basis to ensure continuous process responsibility in the process and an order/deliver relationship between the processes. Designing such a process based organization, with the innovation architecture as a starting point that represents the innovation specific strategic intentions, a cluster of methodological knowledge is a process. At the same time the objects under the methodological knowledge are the input for the process, and the output is the objects above the methodological knowledge. Therefore, such an innovation process model has the same number of cascades as the innovation architecture, and additionally, the segmentation of the innovation processes is the same as for the methodological knowledge. This cascading and segmentation concludes with the different processes to be implemented, which are based on the innovation architecture and highly adapted to the intended innovation opportunities decided in the innovation strategy.[21] Furthermore, the innovation processes are steered by a strategic innovation management process, which is in fact the process on the strategic level of Figure 5.14.

To illustrate this method of designing innovation processes, the case of Pixel AG is briefly presented and shown in Figure 5.34. Due to the fact that Pixel AG has decided to develop a standard full solution product, in addition to mass product development methodological knowledge they require methodological knowledge for developing a standard IT solution. It is obvious that these two mindsets of methodological knowledge are completely different; therefore, the processes should also be different. On the module development cascade, a homepage should be developed that requires a different methodological mindset from the integration and testing of technologies for developing hardware modules. Homepage development needs software programming methodological knowledge. For this purpose also, here the processes are segmented. For the market development as well as for research, segmentation is not necessary because the methodological knowledge will not be different on each cascade. This leads to the macro model of the innovation process of (Figure 5.34), which has to be detailed in further steps.

Control, implement change, update

Based on the innovation strategy and the redesign of the innovation development processes, the implementation can be started. According to Afuah (1998: 217) attention has to be paid to implementation barriers in the specific context of innovation strategy. These implementation barriers can be of an economic nature, (such as cannibalization of existing products, large costs, fear of being stranded), or an organizational nature such as missing capabilities, politic power, emotional attachment to old products, dominant logic, and soon (Afuah, 1998: 217). Optimally, these implementation barriers were all outlined in a previous step.

Therefore, according to Hunger and Wheelen (2002: 121f.), strategy formulation and strategy implementation should thus be considered as two sides of the same coin. To begin the implementation process, strategists must consider three questions:

- Who are the people who will carry out the strategic plan?
- What must be done?
- How are they going to do what is needed?

Management should have addressed these questions, and similar ones, initially when they analyzed the pros and cons of strategic alternatives, but the questions must be addressed again before management can make appropriate implementation plans. Unless top management can answer these basic questions satisfactorily, even the best-planned strategy is unlikely to provide the desired outcome.

To support the implementation successfully, it is also essential to implement an appropriated control system that gives the management feed-back so that it can take updating measures. According to Hunger and Wheelen (2002: 151), a five-step feedback model is essential to design a control system:

1 Determine what to measure. Top managers and operational managers must specify implementation processes and results to be monitored and evaluated. The processes and results must be measurable in a reasonably objective and consistent manner. The focus should be on the most significant elements in a process – the ones that account for the highest proportion of expense or the greatest number of problems. Measurements must be found for all important areas regardless of difficulty.
2 Establish standards of performance. Standards used to measure performance are detailed expressions of strategic objectives. They are measures of acceptable performance results. Each standard usually includes a tolerance range, which defines any acceptable deviations.

Standards can be set not only for final output, but also for intermediate stages of production output.
3 Measure actual performance: Measurements must be made at predetermined times.
4 Compare actual performance with the standard: If the actual performance results are within the desired tolerance range, the measurement process stops here.
5 Take corrective action in terms of update: If the actual results fall outside the desired tolerance range, action must be taken to correct the deviation. The action must not only correct the deviation, but also prevent its recurrence.

Conclusion

The presented innovation strategy formulation process was developed based on the identified gaps in literature as well as in practice. These gaps were identified based on eleven criteria that an innovation strategy formulation process should contain (presented in Chapter 2). Now these criteria should be the basis for validating – from a theoretical point of view – the developed innovation strategy formulation process based on the innovation architecture.

- The innovation strategy formulation process provides a strategic management specific understanding of **complexity**, **systemic interaction** and **evolution**. To provide this understanding, innovation architecture is the basic tool. With its structured visualization of object, methodological and meta-knowledge, the complexity is reduced and the systemic interaction is presented. The evolution is integrated into the innovation architecture by the functions and the innovation fields as well as the object knowledge that does yet not exist.
- The innovation strategy formulation process provides a strategy-specific understanding of **direction, focus, organization** and should be **consistent**. Because the decisions that are taken in the innovation strategy formulation process are based on a high level of understanding of the innovation system, and because the innovation architecture shows all innovation opportunity in detail, it is possible to check the consistency, define clear objectives and set the path by giving a clear direction. Additionally the focus can be defined in the innovation architecture and an adequate organization can be derived.

- The innovation strategy formulation process provides an innovation-specific understanding of **integral innovations, innovation barriers, innovation newness** and **innovation-relevant knowledge**. Because the **process** is based on innovation architecture, the innovation-relevant knowledge is integrated. The integral innovations are ensured due to the fact that business and technological innovations are mentioned in the innovation architecture directly, and organizational innovations are derived in the roll out step. The newness of innovation is integrated by its level of knowledge for each object. The innovation barriers to consider are also integrated indirectly because the strategic decision is prepared in a holistic manner by using many tools for different purposes. These tools allow the detection of many of the potential innovation barriers.

In a nutshell, the innovation strategy formulation process seems, from a theoretical point of view, to be a solution for closing the gaps presented in Chapter 4. But from a practical point of view, the innovation strategy formulation process should be evaluated in terms of its practicality and its implementation based on action research. This is the subject of the next chapter.

6
Action Research

In this chapter, nine action research cases[1] are presented showing the implementation of the developed innovation strategy formulation process, including the innovation architecture (see Chapter 5). These action research cases were accomplished through consulting projects and academic industry projects. The focus of the action research cases is first of all on actively analyzing the practicality of the concept by implementing the concept. Additionally, documenting the modelling procedure of the innovation architecture and innovation strategy formulation process in terms of a practical handbook is an objective. Furthermore, the developed concept could be validated by an individual analysis and a cross-case comparison. This allows a conclusion to be derived and the working hypothesis to be rethought.

In the following action research cases, the company names are changed and data is abstracted because of two factors. Firstly, including all the details would go beyond the scope of this work. Secondly, the results of the cases are very sensitive regarding the content of the collaborating companies; changing the companies' names allows focus on the conceptual elements and results. Nevertheless, the essential aspects of retracing the implementation of the concepts are not omitted.

Selection and procedure of the action research cases

To verify and analyze the usefulness of the developed concept of innovation architecture and innovation strategy formulation process, enterprises or organizational entities were selected that are different from each other and cover a representative range of diverse industries. These enterprises and organizational entities differ, for example, in their industry affiliation, size (sales, employees) and financial success (income in percentage of sales) (see Figure 6.1).

		Industry affiliation	Size		Income
			Sales	Employees	(% of sales)
1	Toll revenue	Component producer	2 bn CHF	8,000	100 mn CHF (5 per cent)
2	TecChem	Chemistry	8 bn CHF	20,000	1.2 bn CHF (15 per cent)
3	Hightec	Production system	1.5 bn CHF	6,000	80 mn CHF (5.3 per cent)
4	Info exchange	IT provider	700 mn CHF	2,000	70 mn CHF (10 per cent)
5	Optic dye	Chemistry	n.a.[1]	50	n.a.[1]
6	Built-up	Apparatus producer	3 bn CHF	15,000	30 mn CHF (1 per cent)
7	Rubtec	Component producer	350 mn $	800	10 mn $ (2.9 per cent)
8	MicroSys	Apparatus producer	600 mn €	4,000	n.a.[2]
9	Stocktec	Apparatus producer	50 mn €	430	1.2 mn € (2.4 per cent)

(1): The company was newly founded in the year of the action research project. No financial data are available.
(2): The company does not make the financial key figures available to the public.

Figure 6.1 The nine action research cases at a glance[2]

The action research cases had a duration ranging from two intensive weeks to three months at a lower intensity. Numerous workshops, interviews, free access to many employees of all the hierarchical levels, analysis of previous innovation activities were allowed to make a detailed and holistic analysis of the cases. The processing of the action research cases, which is shown in the following sections, is divided into five blocks:

- *Short introduction of the companies* By a short introduction of each company, the environment of the organizational entity or company can be presented. On the basis of general but important key figures, it would therefore be possible for the interested reader to understand the situation of the company and to compare it, eventually, to their own company's situation.
- *Initial position* In the 'initial position' block, the company is analyzed in the context of innovation. To analyze the company a set of criteria is chosen to show the important factors in terms of innova-

tion strategy formulation,[3] strategic innovation structures, strategic innovation objectives, strategic innovation behaviours[4] and innovation decision processes.[5] This set of criteria represents the underlying structure in the interviews. This procedure allows the initial position of the project to be demonstrated and the reasoning of the chosen project objectives to be illustrated.
- *The innovation architecture* On the basis of nine selected action research cases (cf. Figure 6.1), the development of an innovation architecture will be described in detail.
- *The innovation strategy formulation process* In every action research case, the company specific modules of the innovation strategy formulation were selected and implemented in the specific company context. These modules were then used with actual strategic innovation-relevant concerns. This implementation is described in detail.
- *Conclusion* The conclusion shows a first appraisal of the innovation architecture and the innovation strategy formulation process. Through this the management feedback is mainly presented.

The action research cases will be presented in the next sections and structured according to the above five blocks. The cross-case analysis and conclusion closes this chapter.

Case 1: Toll Revenue

Short introduction

This action research case was done in a division of a globally active corporate group. The group is active in the telecommunication, automation and energy systems segments. The group employs a total of 8000 employees. The automation sector, in which the division 'Toll Revenue' is embedded, provides in the service sector products, systems and services for the rationalization and automation of routine activities in the service sector, such as issuing entrance cards. The focus of this segment was, near the extension of the market position and restructuring of the organization, to integrate E-solutions into the product portfolio.

Toll Revenue offers electrical apparatus to collect fees. The clientele comprises small, medium and large scaled enterprises as well as government agencies. The largest business volume is reached in the markets in Northern America, the United Kingdom, Germany, Switzerland, South Africa and Scandinavia. The sales figures decreased from 2001 to 2002 by 13 per cent after a long period of constantly increasing sales. The result was that the loss increased from –6.5 per cent to –10 per cent of sales. The reaction to this negative trend was that in 2000/01 a comprehensive reorganization of the division was undertaken, which, in the short term, had a positive effect on costs.

The product portfolio of Toll Revenue is actually affected by a major change. In the past, the products were sold as stand alone solutions, mostly based on hardware developments, but in the future the essential element to gain competitive advantage will be to integrate the products into a communicating system, which is much more software based. This change was not optimally accomplished by the division because important products had delays in development and, therefore, in market introduction.

The R&D budget of Toll Revenue was about 30 mn CHF (10 per cent) and in the future it will decrease.

Initial position

Strategic innovation structures: The development of new products is decentralized into two major locations. These two locations are independently developing new products without harmonizing their objectives. Important overlaps in their activities are the result. In 2001, a reorganization of the innovation system was implemented to increase the effectiveness and efficiency in the two locations, and also better to coordinate the activities more effectively. The coordination of this reorganizations has not yet shown major effects because of structural political barriers.

Strategic innovation behaviours: The management of Toll Revenue, because of the poor financial situation, is very short-term oriented. Therefore, the R&D budget was reduced and only short-term effective projects are allowed. The communication between strategic management and operational developer is reduced to these short-term projects. This effect is aggravated by the fact that the market side management is not coordinated with the technology side management for mid- and long-term product and technology projects. There is an important lack of transparency over the innovation system.

Strategic innovation objectives: The main innovation objective is to cut costs in development and in actual projects, and not develop innovative products. To do this, strategic management wants to have a transparent overview of the development activities on technology and product levels.

Innovation decision processes: The actual strategic decision process is very short-term market oriented, which is the result of the financial situation. Therefore, mainly regional and product marketing is involved, with the executive board deciding on the different projects. R&D is mainly involved passively in this process, to define the needed resources for the proposed projects of the marketing section. R&D's passive role in the innovation decision process is not optimal due to the fact that long- and mid-term projects based on technologies are not integrated into the process. Therefore, a major innovation step could not be realized.

Project objectives: To pull together the situation of the innovation decision process, a two step procedure was decided. Firstly, the co-ordination of actual and future innovation activities should be shown in a transparent overview by the innovation architecture in order to detail technology push and market pull activities, and to show the real strengths and weaknesses of the whole innovation system. In a second step, these activities should be evaluated in terms of timing, feasibility and resource allocation.

The innovation architecture

The innovation architecture of Toll Revenue (see Figure 6.3) was developed with the focus of achieving an overview of the actual innovation activities and ideas, and also subsequently evaluating the object knowledge. The innovation architecture mainly presents the object dimension, partly the methodological knowledge dimension and omits the meta-knowledge dimension.

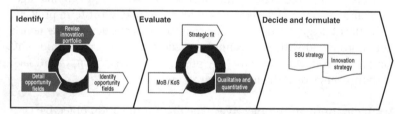

Figure 6.2   Project objectives of action research case at Toll Revenue

Four cascades were identified. On the basic cascade there is externally acquired, applied knowledge developed into technologies. These technologies are categorized into six strategic technology platforms.[6] On a second cascade, These technology platforms fulfil the six functions out of which the modules of the company are developed. The functions in the innovation architecture are defined based on the product function, which is primarily the identification of a communication need of the customer. Afterwards, it is the proposition of a service that the customer wants. Thirdly, money has to be transferred and fourthly, the service is offered. Additionally, the product of Toll Revenue can, if desired, offer additional services other than collecting tolls, and product functions to operate the devices. These functions that are fulfilled by technologies are developed – on a second cascade – into the modules, which are integrated into the whole product/service on the third cascade. The fourth cascade is responsible for transferring the products to the customer and developing marketing concepts. For each cascade the specific methodological knowledge required is presented in Figure 6.3.

The innovation architecture shows the different levels of object knowledge for each object. Thereby, a categorization of five different levels is created: 'existing', 'planned to be available', 'not available', 'partly existing' and 'needs to be updated'. These five levels help to understand the actual position of the innovation system at Toll Revenue.

The innovation fields of the innovation architecture are 'ticketing', based on the fact that today all products of the company offer a service which is confirmed by a ticket and 'system management', based on the fact that all products of Toll Revenue are connected with highly complex system management tools.

The innovation strategy formulation process

Identify

In this step, the innovation architecture was developed and subsequently detailed. This procedure was performed in a cyclical process.

During the innovation architecture based revision of the innovation portfolio and detailing of the innovation opportunities, three important weaknesses of the innovation system of Toll Revenue were found:

- Most planned innovation activities of Toll Revenue are based on activities of existing products which are adapted or changed.

136 Structured Creativity

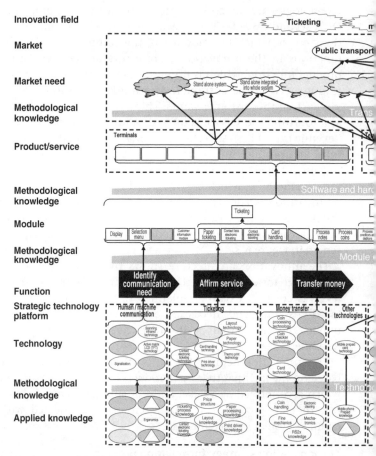

Figure 6.3 Innovation architecture at Toll Revenue

- The innovation activities that are based on a technology push in the six strategic technology platforms are almost not linked to the satisfaction of the customer needs identified by marketing. For example, the products 'other ticket machines', 'information systems' and 'P&R systems' do not even have a link to modules, functions or technologies as presented in the innovation architecture of Figure 6.3. At the same time, the few linked technologies are only linked to existing and traditional modules and products, which have a tendency to decreasing orders.

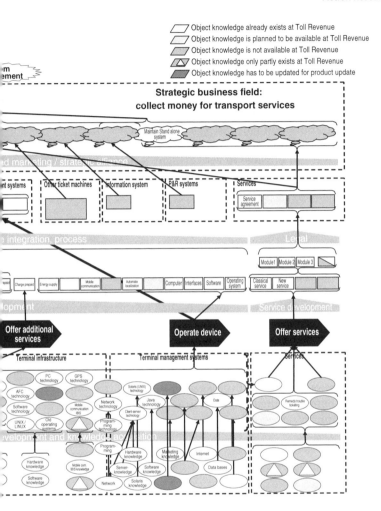

- The function 'offer additional services' is, according to the statement of the head of innovation and marketing, a highly increasing domain, but at the same time innovation activities are very rare.

Concluding the innovation architecture process helped the management in the identification step firstly to acquire an overview of the activities in the dramatically unstructured innovation system. Secondly, the awareness of the missing functional handshake between technology push and market pull was obviously visible, and thirdly, the management saw that the future of the company was not aligned to the customer needs.

Evaluate

In the evaluation, the objective was to make a qualitative and quantitative evaluation of the innovation system. The aim was mainly to develop an innovation roadmap and an resource allocation optional. Due to the dramatic unstructured situation of the innovation system of Revenue Toll it was not possible to get key figures about the duration of several projects, nor was it possible to get key figures about the costs. Therefore, the project aim to make an evaluation was abandoned.

Conclusion

Although that the evaluation step could not be undertaken, the management concluded that the innovation architecture showed the actual situation of Toll Revenue very well. They concluded that the situation was more dramatic than they had previously thought. Based on these newly gained insights, the management decided firstly, to formulate a new corporate or business unit strategy, which was actually missing and therefore resulted in an important lack of coordination. In a second step, the innovation architecture was to be aligned to this new strategy. Only after this step would a second possible evaluation step of the individual innovation opportunities seem to be realistic. Additionally, the management defined the need to sample more data in order to develop an innovation roadmap.

Case 2: TecChem

Short introduction

The TecChem action research case was worked out in a business segment of a worldwide chemicals producer. This company employs about 20 000 employees. Every business segment is responsible for its own sales, marketing, production, development and partly for research.

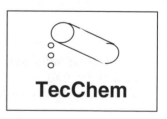

Some research projects, based on core technologies, interesting for the whole company, are done at a corporate level. The focus of all the activities in the different segments is based on the simplified strategy: 'Expansion through innovative technologies'.

The specific business segment, TecChem, offers molecules for producing layers with different functions. In this market, the segment has a relative income of more than 20 per cent of sales, which is above the industry average. This high income results from older products, in terms of 'cash cows',[7] that have not yet been copied by competitors. But, at this time, potential new technologies will allow the same functions to be developed with a higher performance. These technologies could provoke a substitute of the actual 'cash cows' of the company.

Initial position

Strategic innovation structures: The structures in innovation are mainly in alignment with the overall business segment structures. Therefore, a very strict market orientation is present. For new emerging and high risk technologies that are important for the whole company, special technology based group research projects were initiated. The head of these group research projects is the CTO who has to coordinate the company activities with the different business segments.

Strategic innovation behaviours: The company has engaged respected chemists with considerable international reputations in different technology domains. R&D is therefore highly driven by these experts. For the company, this expert driven development is an advantage, as long a product is based on one of the technologies mastered by one of the experts. However, in the event it is not, the company reacts very slowly because of a lack in knowledge. Another important point to mention is the strong and effective network between the experts in the company. This allows the coordination of complex activities in an uncomplicated and effective manner.

Strategic innovation objectives: Because of the experts, who are specialized in a specific technology, the company wants to base their innovation activities on these technological domains. The strategic innovation objective is primary to innovation, by leveraging actual technologies for specific market needs. Nevertheless, the company is aware of the need to monitor new emerging technologies.

Innovation decision processes: The innovation decision process of TecChem is dependent on the type of the intended project. If the project is located in a business segment, the identification is done by the segment experts. In the event, the evaluation is based very strongly on market evaluation tools. If the project is part of a research group the

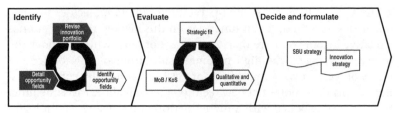

Figure 6.4 Project objectives of action research case at TecChem

identification is mainly based on experts' ideas or general technological trends, and the evaluation is instead based on technological criteria, such as the future importance of the technology and the ability to leverage it in the company.

Project objectives: Because the innovation decision processes on strategic evaluation are already very sophisticated, the management did not have a particular interest in new evaluation tools. Nevertheless, they wanted a tool to identify existing in-house competencies. It was therefore decided to construct the innovation architecture. This would give the CTO the opportunity to have an overview of actual competencies and their interaction for use in preparing decisions for future innovation activities. Furthermore, the elaborated innovation architecture could be used in future as a communication tool for the different business segments.

The innovation architecture

A simplified innovation architecture for TecChem is presented in Figure 6.5. The innovation architecture is simplified due to the fact that, firstly, TecChem has many secret long-term innovation activities shown in the original innovation architecture and, secondly, because the original innovation architecture was very extensive. Nevertheless, the essential elements are seen in the innovation architecture or will be described.

During the project, five cascades were identified in the innovation system at TecChem. The first cascade basically searches in 'scientific knowledge', searching for new insights about molecules in several substance classes. This knowledge on molecules is used to design and synthesize molecules with an effect that could be of interest in the future. These effects, which are summarized in the functions (10 functions in the case of TecChem) are leveraged in a next cascade for developing specific products and scaled later up on another cascade. At the top of the innovation architecture, these products are brought to the market by the application marketing cascade.

Based on the project objectives, the innovation architecture should serve as a visualization and communication tool. Therefore, the different objects were clustered by their departments in order to see the responsibilities. Additionally, it was not possible to present the extens-

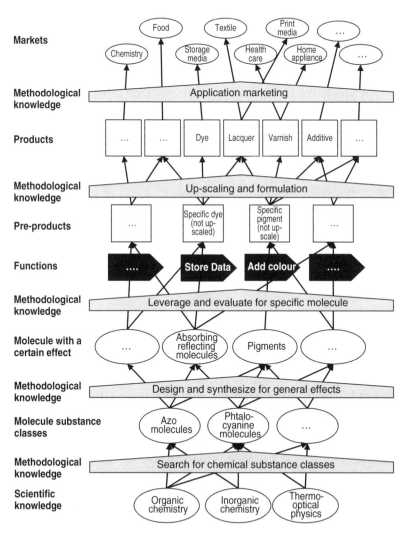

Figure 6.5 The innovation architecture at TecChem (simplified)

ive innovation architecture on one overall paper. Therefore, the visualization of the innovation architecture was done with hyperlinks, to zoom directly in one section.

The innovation strategy formulation process

Identify

During the 'identify' step, several interviews were conducted throughout TecChem. During these interviews it was quite easy to integrate the objects into the innovation architecture. Also, the links between the different objects could be integrated in very quickly. However, during the detailing of the innovation opportunities it appeared that the allocation of responsibilities between departments was not clear. This relates to the fact that some researchers were active in the cascade of searching chemical substance classes, which is solution neutral, and simultaneously to the cascade of leveraging specific modules, which is solution oriented. The disadvantage of this fact was that the research became solution oriented or that the development of products was not primarily based on a specific customer need but on an effect that was discovered in research.

Additionally, it must be mentioned that during the development of the innovation architecture the employees in the innovation system were somewhat astonished to learn about the project and competencies of their colleagues. Therefore, the innovation architecture helped to align the objectives of the departments.

Conclusion

The innovation architecture of TecChem was accepted in the main by the management to visualize the knowledge that exists in-house and to better allocate responsibilities. Additionally, the innovation architecture allowed weaknesses in the innovation system to be visualized, such as the unclear allocation of responsibilities. Although the innovation architecture revealed some interesting points in the company, the management criticized the concept of innovation architecture. In the opinion of the management, the effort to build up the innovation architecture was too great, compared to the results.

A major cause of difficulty in is trying to explain the benefit to these critics that the innovation architecture was only used as visualization and communication tool, not as a tool to identify, evaluate or derive organizational processes. Owing to this limited use of innovation architecture, the input of effort was very high. This action research case therefore showed that innovation architecture is a not a tool that should simply be used as a communication and visualization tool. Innovation architecture is more a tool for identifying, evaluating and deriving organizational processes, although an interesting side effect is that it is a good basis for the visualization of innovation activities and competencies.

Case 3: HighTec

Short introduction

HighTec, is a leading provider of production systems, components and services for selected growth markets, focused on information technology and sophisticated technological applications. It employs approximately 6000 people and achieved sales of about CHF 1.5 bn with a loss of about 5 per cent of sales. As a production system provider, the company is at the beginning of the value chain in this industry. Due to the bullwhip effect,[8] this results in sales figures that are highly unstable. In addition to this effect, the high tech industry is by nature very unstable and dynamic. Therefore, HighTec sales can vary up to 50 per cent in a year. This is one of the major challenges for the company to master.

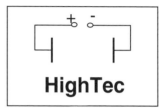

Two of seven independent divisions of HighTec were analyzed in terms of action research. Both analyzed divisions, each divided in several strategic business units (SBUs), are active in the same industry, but with different markets and products. Both divisions basically do not vary in employees (both about 500), in sales (both about CHF 200 mn) or in the challenges their need to master. Therefore, the introduction and initial position is common for both divisions.

In the last few years, sales were constantly growing but broke down one year before the action research case. Due to this effect, many problems that had not been seen in times of high sales began to occur. For example, a major concern was that new products had problems with quality, and therefore an R&D team had been working for a long time for a customer. In good times these costs were not seen as important, but nowadays these costs, compared to sales, form a high value of fixed costs that cannot be reduced quickly. These quality problems barely affect R&D, because resources are blocked and new product launches are delayed (the R&D budget is 10 per cent of sales).

At the time the action research project started, the company-wide strategy was in formulation. The main question to be answered in this process was: 'Is it better to focus on standard production systems or on customized production systems for specific client needs in market niches?'

Initial position

Strategic innovation structures: R&D is organized in its structure, so that innovation projects are done by one team. This team has the task of undertaking research, technology development as well as product development. This organization allows a highly flexible structure in terms of resources but, however, the projects are extensive, long and change frequently over time as customers' needs change. Therefore, this structure of R&D needs a high level of transparency and considerable effort in coordinating these projects. However, this transparency and coordination is often lacking.

Strategic innovation behaviours: Because the projects are often extensive, the project leaders have no time to coordinate their activities with other SBUs, a situation often not gives consideration at any level in the company. Furthermore, the coordination between development activities and market introduction is not optimal, which results in quality problems. In addition, the strategic mid- and long-term coordination in place is only rudimentary.

Strategic innovation objectives: Due to the fact that the sales decreased in the last year, the strategy in innovation is mainly to strengthen the innovativeness, to implement a strategic steering, focus resources and focus on attractive customer needs with the main objective being to increase market share.

Innovation decision processes: The innovation decision process is, in theory, aligned and therefore a sub-point of the general strategic decision process. The general strategic decision process is based on a three-year strategy, which is detailed every year by the different SBUs and accepted by the management to make an annual plan.

Project objectives: To support the general strategy process in terms of innovation-relevant issues, the following objectives were defined: (1) revise and detail the innovation portfolio on the basis of the innovation architecture, (2) conduct a quantitative and qualitative analysis

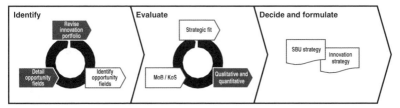

Figure 6.6 Project objectives of action research case at HighTec

in which a roadmap is determined to identify bottlenecks in the actual projects in term of resources and time (Figure 6.6). Based on the fact that the two divisions of HighTec analyzed are highly independent, the project was performed autonomously in each division.

The innovation architecture

Based on the fact that the two divisions were separate, the different circumstances of the two projects led to two different visualizations of the innovation architecture, as can be seen in Figure 6.7 for division one and in Figure 6.8 for division two. Therefore, in the following section the two innovation architectures are described separately.

Innovation architecture of division one

At the beginning of the process of developing the innovation architecture it was clear that the company had a small number of products and modules compared to technologies. Therefore, a visualization of the innovation architecture (see Figure 6.7) was chosen to show technology and science on different slides, which are segmented by the function they fulfil.

The elaborated innovation architecture has five cascades. The lowest cascade consists mainly of research knowledge, which has to integrate new scientific insights into basic technologies. This research cascade only existed in four of the 12 segmented slides, because, according to the management, only there is it essential to do research. The basic technologies are then developed by the applied technology cascade, with the aim of developing concrete technologies that fulfil a specific function (such as, for example, the technology 'sputtering' fulfils the function to 'depose material'). These functional based technologies are then developed on another cascade into modules, which are subsequently integrated into a complete product line based on the methodological knowledge of the product line development cascade. This product development is, in this case, especially important because its

146 Structured Creativity

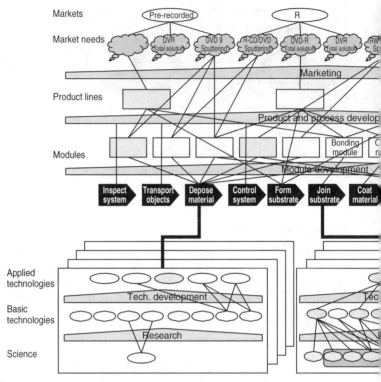

Figure 6.7 Innovation architecture at HighTec (Division One): segmentation by function

main methodological knowledge consists of how to develop a product line architecture. At the top, the marketing cascade is charged with preparing the market introduction.

The objects were clustered into three knowledge levels (exists, planned, not available) and additionally it was mentioned that some objects are outsourced.

Innovation architecture of division two

The products of division two are marked by the fact that they are complex and extensive. Mainly, this is because they consist of many different modules that are not leveraged throughout the products. This fact led to a visualization of the innovation architecture where the slides were segmented by products in order to ensure a focused and not too complex overview (see Figure 6.8).

Action Research

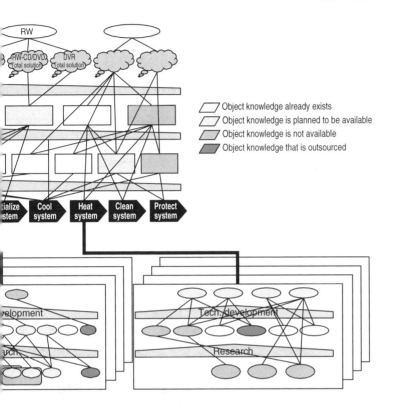

The innovation architecture of division two consists of four cascades. The cascade on the bottom merges with the research and technology development cascade, in contrast to the case of division one. This merging into one cascade is based on the fact that research in this division does not play a major role. On the next cascade a two-fold segmentation of the methodological knowledge was done. Whereas one segment consists of methodological knowledge to develop an autonomous module, the other segment consists of methodological knowledge to develop the handling system, which can be seen as the basic platform for all the modules. The cascade above these two segments is the solution development for developing the customer specific products and at the top the market introduction is prepared on a separate cascade.

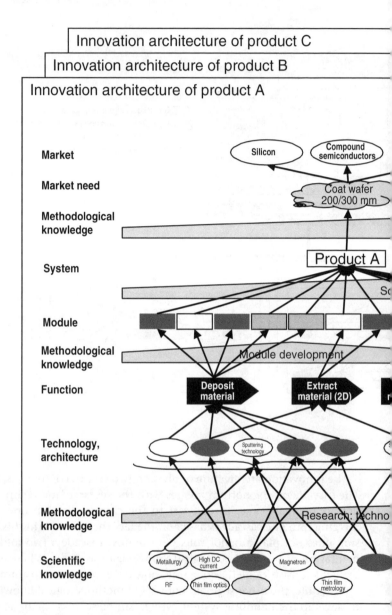

Figure 6.8 Innovation architecture at HighTec (Division Two): segmentation by products

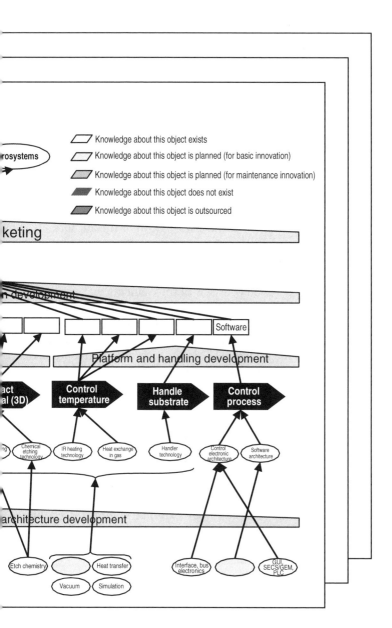

The levels of knowledge were differentiated by four categories (exists, planned, not available, maintenance needed). Additionally to these four categories, some objects are outsourced.

The innovation strategy formulation process

Identify

In both divisions is important to mention that during the process of revising the innovation portfolio and detailing the opportunity fields, there was no problem in defining all the objects needed to develop a certain innovation opportunity. Therefore, the innovation architecture was constructed in a very short time when compared to other projects. One reason for this phenomenon – in the opinion of the authors – is the fact that the whole innovation system is vertically organized. This means that an innovation opportunity is mainly developed from scientific research to market introduction by one project group. This group therefore has a very good overview of the elements that are missing. But, at the same time, in the innovation architectures of both divisions it could be identified that technologies and scientific knowledge were in fact integrated, but there were a tendency that there was a low level of knowledge for the integrated objects. According to the project manager, this lack occurs due to the fact that the project members were at the end of a product introduction and focused on perfection of these products due to quality problems. But, because of the vertical organization, these project members could not concentrate on technology development and research for the next product generation.

Another aspect that was found during the process of architecting was the fact, that – especially in division one – all the technologies, which were needed to fulfil the functions, were developed internally. Even the function 'cool system' was developed mainly in-house, whether or not it was known in the division that external companies could develop this cooling system better and cheaper. This is one reason that division one has such a large number of technologies, which is difficult to manage.

In division two, the innovation architecture was segmented by its products because of the different modules. But at the same time, the different innovation architectures showed that although the modules are different, the technologies are often the same for all the products. After a more detailed study, it became evident that the products had different modules not because they needed to, but because a product was always completely redesigned. However, some managers said that this was not necessary. Due to this fact, the product complexity was enormous and difficult to handle.

Evaluate

Based on the innovation architectures of both divisions, innovation roadmaps were generated.[9] The innovation roadmaps were developed from a market viewpoint. This means that, firstly, the market introduction date was fixed and the projects were in a second step back in planning. The result was that both divisions could not finish their innovation projects until the fixed market introduction date. The reason for this lack of planning was different for both divisions as now described:

In division one, the identified lack of object knowledge in technology and science had dramatic impacts on the innovation roadmap (see Figure 6.9) of which management was unaware. In one case, a certain technology had to be available in-house from the end of 2002 to begin the module development. But in reality, with the given resources the technology could only be available from mid-2003, which should have been the deadline for the module development. However, market pressure demanded that the planned deadline of the module development should actually be the date of its market introduction. The company would have only finished the technological development by the due date for market introduction. Management knew that a delay was highly probably but not that it would be so dramatic.

The innovation roadmap of division two (see Figure 6.10) also showed that a technological knowledge gap would provoke a delay without changing the resources, but in their case this was not the major problem. More dramatic was, firstly, that the management thought that a specific module could be developed in-house, but the project members knew that this was not possible. Secondly, an old product, named in Figure 6.10 (product A (V1.0)) had quality problems which had to be solved. But the resources used to solve these problems could not develop new products. However, the innovation roadmap, and therefore the time planning of the project leader, was done in the belief that they had all resources available. Therefore, the delays would increase compared to the innovation roadmap in Figure 6.10.

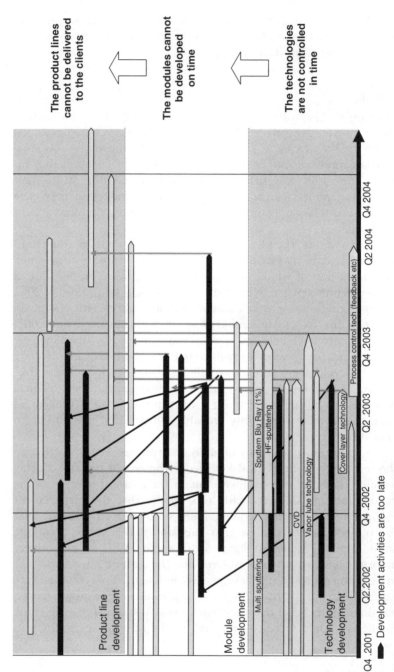

Figure 6.9 Innovation roadmap at HighTec (division one)

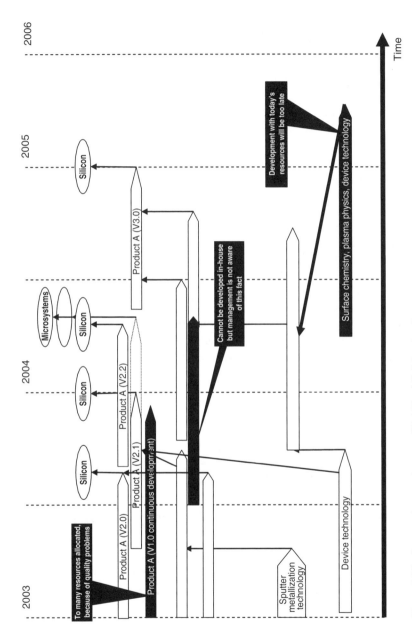

Figure 6.10 Innovation roadmap at HighTec (division two)

Conclusion

To conclude, the process of architecting helped in this specific action research case to find out the following aspects, as affirmed by management:

- The knowledge gap, especially on the technological side, turned out to be massive.
- In division one, the number of technologies to be developed in-house was very high, even though it was known that external partners could do it better.
- In division two, the number of modules was much higher than it needed to be.
- Neither division had enough resources for all the development projects. The resource planning had not been done systematically.
- The innovation roadmap showed that the coordination, in terms of timing, between the market side and the technology side was dramatically misaligned.

The management said that this innovation architecture, as basis for the strategy process, had helped to reveal several problems that can now be described based on facts. It was also mentioned that these problems were discovered especially by the process of architecting, which forces the right questions to be raised and answered.

Case 4: Info Exchange

Short introduction

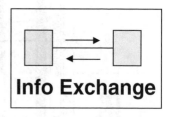

The division analyzed in this action research case is part of a company active in only one country. The Info Exchange provides services and products in the areas of international financial information, cashless payment instruments, electronic payment systems and IT services. All these activities are based in the business of providing specified IT systems with a high level of security and availability. About 2000 employees are employed and the sales had been stable at 10 per cent of sales in the last three years.

The specific division of 'Info Exchange' is the internal IT provider for the other divisions. These internal orders comprise about 88 per cent of total sales, and they are the exclusive IT-providers to the other divisions. Therefore, they can act as monopolists in an ideal market. But this ideal situation will change due to a strategic decision: management decided that the division had to increase their sales by doubling sales to external clients from 12 per cent to 24 per cent of total sales.

This challenge to double external sales is complicated by the trend that the trust in IT-products is in general decreasing, originating from a poor level of service quality and insolvency in the IT industry. This trend provokes a tendency to loss of sales by 10 to 15 per cent by year. To react to this situation, the aim of the division is to gain the trust of the customers by focusing on core competencies to develop new products with a high quality in new attractive markets.

Initial position

Strategic innovation structures: The hierarchy is flat which allows, in innovation, an optimal lead and fast information flow. This flat hierarchy underlies a newly implemented process organization where a CTO (Chief Technology Officer) has the task of coordinating the innovation activities division-wide.

Strategic innovation behaviours: The innovation culture is, due to the fact that the company could act for a long time as a monopolist, not very innovation friendly. Additionally, the resources for new investments are handled very restrictively. Therefore, only innovation projects are accepted that are based on a specific customer demand. The technology department is not asked about future possibilities for new functions but only about the feasibility of the demanded innovations.

Strategic innovation objectives: Based on the division strategy mentioned above, the rough innovation strategy has the aim of developing innovations for growth based on actual core competencies deployed in new markets. For this very rough innovation strategy, there was not yet a defined path to reach the objectives. The technological core competencies were not defined with a focus for identifying new potential markets.

Innovation decision processes: The innovation decision process is implicit. Based on the initiative of corporate management, the divisions have to formulate a realistic strategy containing mainly objectives but not paths. This development is based on financial key figures. A forwarding systematic identification phase is missing. The formu-

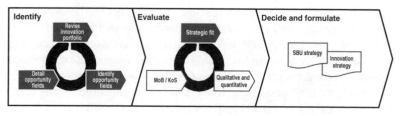

Figure 6.11 Project objectives of action research case at Info Exchange

lated strategy is then consolidated on a company level. A specific and detailed innovation strategy does not exist.

Project objectives: The project objectives were concentrated on the identification of new markets based on existing innovation activities. Therefore, in a first step the existing innovation activities, including strengths and weaknesses, should be defined in the innovation architecture by revising the innovation portfolio. In a second step, innovation opportunities in new markets should be identified, based on related activities, and detailed. In a third step, a preliminary strategic fit evaluation should be done (see Figure 6.11).

The innovation architecture

The innovation architecture of 'Info Exchange' (see Figure 6.12) has four cascades. On the first cascade, the company needs methodological knowledge to screen for new technologies. Because Info Exchange is not developing new technologies, it is a very important element during the development of any new product to be up to date in a rapid changing environment. This technology screening is done for six strategic technology platforms (STPs) that were defined during development of the innovation architecture. These STPs have the aim of fulfilling the functions, which are needed in separated modules. This development of the modules, the next cascade, is segmented in five different methodological segments because of the fact that, for example, hardware and software design are completely different in their methodological knowledge. All developed modules consisting of hardware and software are integrated in the next cascade to a customer specific product or service. This product is then introduced into the different markets on the basis of the next cascade.

In addition to the innovation architecture of Figure 6.12, that represents the object and methodological knowledge dimensions, the meta-knowledge dimension was analyzed also in its five categories.[10] In the following, an example of Info Exchange's general meta-knowledge is shown:

- The source of knowledge is based on internal employees, external education and a personal network.
- The insurance of the reliance of knowledge is based on statements of the employees.
- The importance of knowledge is mainly growing in the domain of integration knowledge. Therefore this is not the specific object knowledge but the methodological knowledge, and will be more important.
- The evolution of knowledge will be ensured by having active contacts with suppliers, universities and experts.
- The cognitive capabilities to develop new knowledge in the company are highly developed.

The innovation strategy formulation process
Identify
Constructing the innovation architecture at 'Info Exchange' uncovered some interesting aspects. The integration of technology and scientific insights presented no big problem, but the integration of future modules and products was more difficult, because the company did not have a detailed overview over the markets. 'Info Exchange' lacked information on their future products. This lack of market information is based on the fact that the company had, in the past, only a limited group of customers. Therefore, the products were defined by these customers and future market development was not seen as important. Nowadays, where the company has to develop products for other potential customers, this lack of market information is a major disadvantage. An additional interesting point discovered during the process of architecting was the small number of new ideas on the technological side. Under more detailed observation, it came out that in the main technologies are developed for a specific product that has been ordered by a customer. Therefore, in the past 'Info Exchange' was not directed by a technology push but mainly by market pull activities.

The identification of new business fields was based on innovation fields. Therefore, a search was made to identify business fields, especially markets that could have an interest in the innovation field 'disaster precaution' in the special context of IT. It came out in this creative workshop that companies in the media, building, IT, health

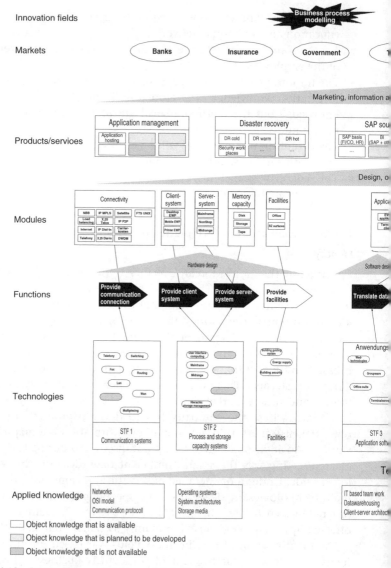

Figure 6.12 Innovation architecture at Info Exchange

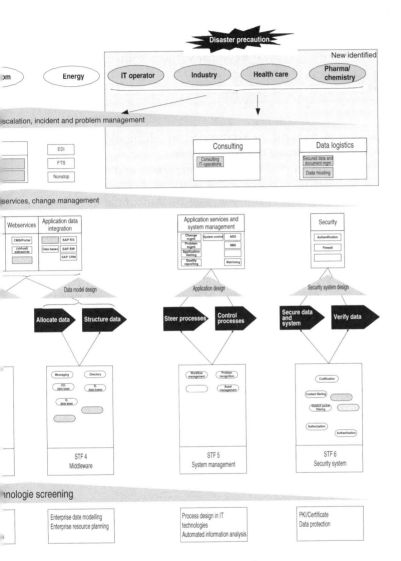

care, pharmaceuticals, chemistry, automobile, logistics, mechanical and retail areas could have an interest. These potential business fields were the basis for the following evaluation. Here, it must be mentioned that the identification of new activity fields, based on specific functions, was not done, because the company strategic direction required that only activities in business fields should be supported where nearly all the existing functions of 'Info Exchange' can be used.

Evaluate

The evaluation focused on the strategic fit which should ensure that only new business fields are interesting when they are based on existing functions in terms of a functional handshake and existing core competencies. Therefore, all the above mentioned potential business fields were analyzed to evaluate how well the functional fit between technology platform and business was accomplished. After this first functional fit, a second-order fit was conducted by identifying whether the company has the competencies in each function for developing future innovations in the specific business fields. The results, which were calculated based on a catalogue of criteria, were summarized for each new business field in its functional potential. For completing the strategic fit, a third-order fit was made by evaluating whether the environment would change positively in the future. Therefore, for each function in each potential business field, the functional attractiveness based on several criteria (such as the market growth or the market size) was analyzed.

These different strategic fits, always based on a functional view, were then summarized in a portfolio that offset the functional potential against the functional attractiveness for each business field. The conclusion of the result of this functional portfolio was that the business fields of IT, pharmaceuticals, health care and mechanics were especially interesting.

Conclusion

The feedback from management about the innovation architecture and the process of identification and strategic fit evaluation was highly positive. It was said that this procedure supports the interdisciplinary teamwork in terms of showing the links between technology and the market side. Additionally, the creativity potential is increased. Nevertheless, at the beginning of the project some employees expressed the feeling that the procedure in this project was in some ways theoretic. However, the results were presented to people that were not project members, and no arguments could be found to call the results into question. Therefore, the overall feedback of management was highly positive.

Case 5: Optic Dye

Short introduction

The fifth action research case was undertaken in the small size enterprise (50 employees) 'Optic Dye', active in the optical storage industry. In mid-2002 'Optic Dye' was created on the basis of an insolvent company. In comparison to the old company, that was active in all the domains of optical storage, the new company only focused on research and development for chemical products in the optical storage industry. The products are mainly laser reflecting dyes, process consulting for producing optical storage devices and forms for the injection moulding of an optical storage device, so called stampers.

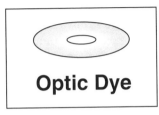

'Optic Dye' has a triple challenge to master. First of all, during the analyzed time the product was still in development, therefore the company had no sales. At all times the investors therefore had to be kept informed. Secondly, the chemical products are already offered, in a weaker performance, by large enterprises that have a very powerful position in the market, and are also developing intensively. Thirdly, the product life cycle is very short (less than one year). These three challenges are forcing the company to develop a higher focus, effective and efficient, combined with aggressive marketing in order to succeed in the future.

Initial position

Strategic innovation structures: Due to the fact that the company is very small, the structures are very flat and implicit. But, nevertheless, there is a CTO in place, who has to coordinate all projects. The CTO, contrary to the CEO, was employed in the old company with technological responsibilities and has a well founded technological background. However, the CTO coordinates all the activities, which makes it difficult to identify the concrete structure of the company.

Strategic innovation behaviours: The employees are very open to internal changes and sensitive to environmental changes. This highly unusual willingness for change could be traced back to the insolvency of the old company. The actual employees have therefore already seen what can happen if internal changes are avoided and environmental changes are ignored. Summarizing, the employees are mainly entrepreneurial, communicative and have a high degree of their own responsibility. This is the great advantage compared to their competitors, which are in general large-sized companies.

Strategic innovation objectives: The strategic objective of the company, concerning innovation, is very focused: be the first in the market for the new laser reflecting dye. To reach this objective the company spotlights one chemical molecule class and is open for strategic alliances with large companies.

Innovation decision processes: An initial unimportant change in the environment could have a major effect on Optic Dye, because the focus of activity is very tight. The executive management is aware of this fact and therefore decides all major technology issues in their executive management meetings. In these meetings, the financial and marketing issues are aligned to the technology based innovation activities. But due to the fact that the CEO has no chemical background, a clear understanding of the technological impacts is difficult for him. However, this understanding, in his mind, is a precondition for taking decisions in a strategic innovation decision process.

Project objectives: The executive management, especially the CEO, of the analyzed company does not have enough insight into R&D activities to take founded decisions. At same time, the management has to react very quickly to environmental changes. For this purpose, the following objectives were defined in a first phase: revise and detail the actual innovation portfolio in terms of transparency with the innova-

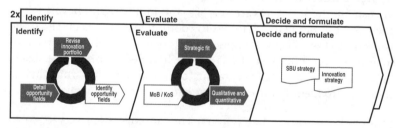

Figure 6.13   Project objectives of action research case at Optic Dye

tion architecture, and introduce tools for strategic as well as qualitative and quantitative evaluation for managing technologies in a small-sized innovation driven company. In a second phase, six months later, the process should be reworked to revise the results (see Figure 6.13).

The innovation architecture

The innovation architecture was constructed in two levels of detail. Firstly, a more general innovation architecture (see top of Figure 6.14) aims to give an overview of how the R&D is structured. This structure shows that the company's activities are based on four strategic technology platforms which are used in several functions. These functions are combined to offer a service (production consulting) and two products (dye and stamper) which are offered to customers in optical storage production. Secondly, a more detailed innovation architecture (see bottom of Figure 6.14) is constructed to support management in understanding the details of the innovation opportunities and their interactions. This innovation architecture has three cascades. The first cascade contains methodological knowledge with the objective of searching for new insights in the domain of science, and using these insights for designing and synthesizing new laser dyes. In the next cascade, the laser dye is up-scaled and the production process is developed. Additionally, on the same cascade, the properties of the laser dyes are documented to gain the know-how of how the dye should be used optimally in the production process of optical storage media. The resulting products are introduced into the market on the basis of a marketing concept. The levels of knowledge were differentiated by three categories (exists, planned, not available).

The innovation strategy formulation process (first procedure)

Identify

The revision of the innovation portfolio and the detailing of all innovation opportunities showed some company specific aspects. The CTO is the person with the most technical knowledge about developing a specific dye. Therefore, the CTO is, due to environmental changes, the only one that has the overview of all of the internal activities. But, at the same time, the CTO is pressured for time, and is highly technical and less management oriented. Therefore, the information about the almost developed objects in the innovation architecture can be identified very quickly and integrated into the innovation architecture. However, due to the lack of the CTO's management orientation, there was no real awareness of how to develop the next generation of dye,

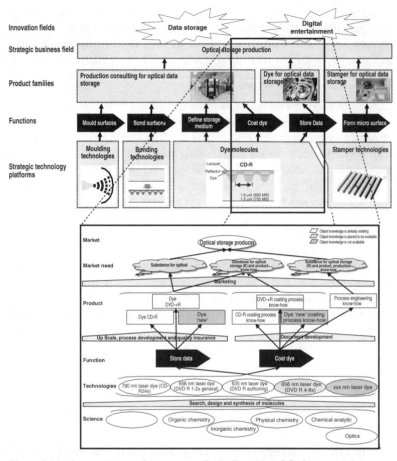

Figure 6.14 Innovation architecture at Optic Dye (simplified overview)

the so-called dye 'New'. Therefore, before completing the innovation architecture, a researcher had to collect detailed information. At the end, the innovation architecture could be developed and presented to the CEO.

Evaluate

The innovation opportunities integrated into the innovation architecture were analyzed, firstly in terms of their strategic fit. Based on the fact that the company is highly focused and that the company has no choice of what innovation opportunity it has to develop (which is defined by the 'big players'), the strategic fit is only made in terms of

165

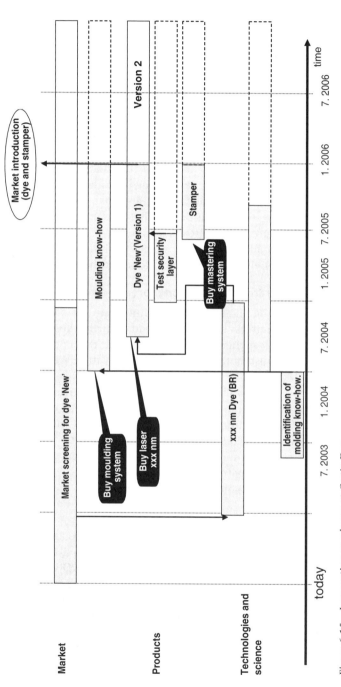

Figure 6.15 Innovation roadmap at Optic Dye

identifying missing elements. The result of this adapted strategic fit was that management saw that in the near future they will need two new machines for developing the dyes and stampers. With these machines, the innovation opportunities could be developed and produced in the future, and therefore a strategic fit is ensured.

Furthermore, a basic quantitative and qualitative analysis was made by using an innovation roadmap. This roadmap (see Figure 6.15) was mainly made for the innovation opportunity of the dye 'New', which is as yet planned. Thus, some additional information had to be collected before transferring the objects of the innovation architecture into the innovation roadmap. During the preparation of this innovation roadmap the most important fact was that the R&D team was sitting together to discuss the time scale. Thus, the required human and financial resources were discussed in detail. According to the CTO, this innovation roadmap meeting was a very important step in synchronizing the individual intentions in terms of the dye 'New'.

The innovation strategy formulation process (second procedure)

Six months after the first innovation architecture was created and the innovation roadmap was approved by management, those results were to be reviewed. It was therefore astonishing that this update was done in half a day because of the fact that the innovation architecture did not change its structure; only the level of knowledge changed and some new objects had to be integrated. Also, the innovation roadmap could be updated easily; only a few objects had to be adapted in terms of time or to be newly integrated. This case therefore explicitly showed that innovation architecture and the tools related to it, due to the innovation strategy formulation process, are – once they are filled with information – very easy to update as long as the basic structure is not changing completely, which is very rarely the case.

Conclusion

Finally, this case showed that innovation architecture and the related tools, adapted in some cases, are also helpful in small-sized innovation driven companies. This is especially the case when R&D is highly technology oriented and not market oriented. Also it must be mentioned that innovation architecture is a tool for communication with top management, to explain constraints between specific technologies and markets, which is often not unimportant. And last, but not least, this case showed that innovation architecture could be updated with a minimum of effort, once the basic structure of the innovation architecture is elaborated.

Case 6: Built-up

Short introduction

The sixth action research case was undertaken in the company Built-up, which is a world leader in developing, manufacturing and marketing added-value for professional customers in the construction industry and building maintenance. The company operates in over 120 countries around the world and employs about 15 000 employees.

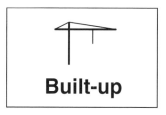

In 2002, the construction industry declined in parallel with the continued weakness of the world economy. Therefore, the company's results fell from an income of 10 per cent of sales to 1 per cent. Nevertheless, the company strengthened its position compared to competitors, according their annual report. The future prognostics for the market development are mostly positive.

The company strategy is to increase the profitability on the basis of innovations (40–60 per cent of sales should be new products of the three last years), operational performance increase and direct market sales. To ensure this corporate strategy in innovation, the company has a growing tendency to invest about 5 per cent of sales in research and development.

The different business areas are independent in their operational activities. In terms of strategic decisions (such as strategy formulation, organizational change or overlapping projects in marketing and technology), corporate management at least takes responsibility for company-wide alignment.

The analyzed business area develops and manufactures tools for the pressurized setting of bolts and nails with efficiency and safety. The business area is one of the traditional business areas active in a matured market.

Initial position

Strategic innovation structures: The analyzed business area is, like the whole company, highly market oriented. Many innovation projects, in particular short-term projects, are based on a specific customer need. Therefore, marketing has a very strong position in the company. Nevertheless, in technological development, especially in corporate technology research and development, in terms of technology push technology based projects are often initiated. These projects are not for a short-term purpose but ensure the innovations in the middle- and the long-term. The strategic innovation structures are therefore market based as well as technology based. To ensure this balanced structure, the company is forced to be organized in a complex matrix organization, with many levels in their hierarchy.

The structure of the processes in the innovation system is divided into search, research, technology development, platform development (optional), product development, product care and phase out processes.

Strategic innovation behaviours: The company can be characterized as a team- and project-oriented company. In these projects, the set of objectives to reach is often very high. Therefore, the employees are strongly performance oriented. Decisions are systematically prepared. The internal employee fluctuation is very high, with a turnover of three years.

Strategic innovation objectives: The strategic innovation objectives are directly aligned to the corporate strategy. The main strategic innovation objective is to increase the percentage of sales of new products. This main objective is divided into coordinated strategic plans for each of the processes of the innovation system.

Innovation decision processes: The strategy process of the analyzed business area, representative for the whole company, is already detailed as soon as the strategic innovation issues are a part of it. In this process, decisions are based on market data (ex. sales, market trends) product data (ex. NPV, cost) as well as technology data (ex. technology trends). This data is systematically evaluated and a decision is taken.

This strategy formulation process is highly sophisticated in terms of evaluation but, due to the complex matrix organization, the foregoing transparency overview of the whole innovation portfolio, essential for decision making, is often only partly present.

Project objectives: To allow the specific business area to get a clear overview of their innovation system, the innovation portfolio is revised and detailed with the innovation architecture (see Figure 6.16). In particular, the link between the market-oriented side and the technology-oriented side should be made visible to show the overall coherence of different activities as part of the innovation portfolio. Thus, management decided, in contrast to all the other cases, to set up the innovation architecture themselves.

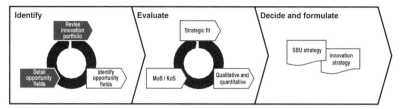

Figure 6.16 Project objectives of action research case at Built-up

The innovation architecture

The innovation architecture of the specific business area of Built-up is based on four cascades (see Figure 6.17). The first cascade, containing general research methodological knowledge (such as FEM analysis) has the aim of searching for scientific insight in order to develop basic technologies for applied scientific effects, which are developed on the next cascade into 'ready to use' technologies. This technology development cascade consists of four different methodological knowledge segments because, for example, the methodological knowledge for mechanical applications is different than for electronic applications. The 'ready to use' technologies, fulfilling the functions of the products, are developed in the next cascade into products, which are introduced into the market with marketing knowledge. This innovation architecture was set up in a very detailed way for the technological side, but not so far the market side. This is because the objective was to show the links between the specific technologies and the general market side. Object knowledge was categorized into three levels (existing, planned to be available, unavailable).

The innovation strategy formulation process

Identify

The revision of the portfolio and the detailing of the innovation opportunities were done completely by management, using the innovation architecture as a basis. Thus, the feedback from management was that they had no problem constructing the innovation architecture, either from a methodological point of view or a content point of view. The content could be easily integrated because the information already existed in the company, which was very well structured in the first place. Therefore, the big benefit of the architecting process was not, as in other cases, to obtain a general overview but rather to identify the important innovation opportunities and create a basis for systematic evaluation.

170 Structured Creativity

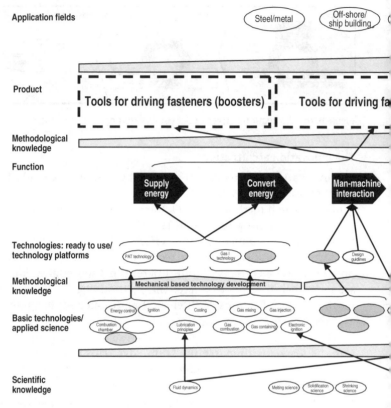

Figure 6.17 Innovation architecture at Built-up

In addition, the definition of the functions was, for the purpose of management, associated with many advantages. Firstly, the functions allowed complex activities to be described by defining their functional objectives, which, secondly, allowed focusing the activities on this main functional objective. Thirdly, the functions allowed discovery, based on a strategic discussion, of a definition of the core functions that define the future direction of innovation activities. Fourthly, the functions are a basis for communication between the market and the technological sides without using a too specific technology or marketing language.

Conclusion

This case showed very impressively that innovation architecture is a tool that management can use without undue studying of the rules, of inno-

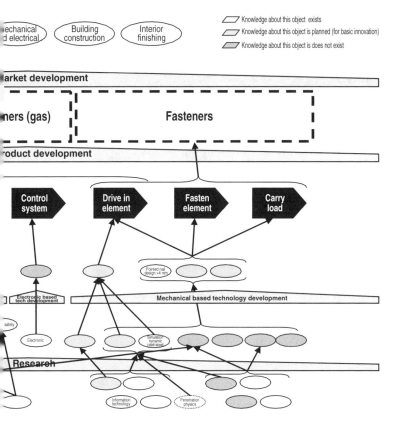

vation architecting. The feedback from management was positive, and expectations in terms of the results were reached. However, due to the fact that the company is already highly structured, the results of the innovation architecture were not high enough in comparison to the required efforts. Nevertheless, due to the advantages of the functional thinking, management was highly satisfied with this action research case.

Additionally it can be said, that this company, which is known in the industry as one of the best in the domain of innovation, has cascaded their processes in the same manner as methodological knowledge is cascaded in the innovation architecture. According to pp. 121ff., the alignment of the methodological knowledge cascades of the innovation architecture to the process cascades is an optimal solution for innovating effective and efficiently. Therefore, this case could be seen as a best practice case for assisting the presented theory of designing an innovation organization based on innovation architecture.

Case 7: RubTec

Short introduction

The Japanese company analyzed in this action research case develops and produces products in several different markets. The important products, in terms of sales, are mainly in the domain of belting, rubber joints, filter systems, mechatronics and sensor systems, and hose and tubing systems. All of these products, with a few exceptions, are in matured markets that are stable or even decreasing. The company employs about 800 employees and is globally active. The income in percentage of sales is about 3 per cent.

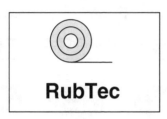

The company, due to its decreasing markets, is forced to identify new attractive markets, Therefore, the company strategy is to strengthen competencies for present products in order to strengthen competencies for new product development.

Initial position

Strategic innovation structures: The company is divided into six independent business segments. These segments are responsible for their sales, production and innovation. Additionally to these segments, a technology centre was formed for developing new businesses or risky projects in existing businesses where the responsibilities are not clearly defined.

Strategic innovation behaviours: The actual important products were very successful in the past, they were 'cash-cows'.[11] Therefore, the need for new products was not recognized. This situation has recently changed. Sales began to stagnate, or even to decrease. This is one of the major reasons that the company today has difficulties mastering entering into new domains. The innovativeness, in terms of being aware of changes, is not developed. This tendency of low innovativeness is enforced, from a European point of view, by a very institutional organization.

Nevertheless, the innovation domain is very effective and efficient. A major reason for this effectiveness and efficiency is, firstly, the very high team orientation in projects, and secondly the practice that new technologies are systematically licensed from external companies. This allows the development of products to be very fast and requires less effort. The 'not invented here syndrome', dose not seem to exist in this company.

Strategic innovation objectives: Based on the above strategy, the strategic innovation objectives are to identify new markets, based on actual competencies, or even to create new competencies for existing or new markets. However, due to the unclear definition of responsibilities between the technical centre and the business segments in terms of development of innovations, the specific strategic tasks to find new markets are not clearly allocated. Therefore, no clear detailed innovation strategy containing objectives and a path is defined.

Innovation decision processes: An implicit strategic decision process exists. But this strategic decision process contains neither the systematic identification of actual activities or new opportunities, nor was a systematic evaluation of these elements found. This is a major weakness on the strategic level, in particular during these times, where the company has to decide about fundamental internal changes and new activities for the future.

Project objectives: Due to the fact that the strategic innovation decision process only implicitly exists, the objective of the project is to implement a process for revising the innovation portfolio, systematically identifying the innovation opportunities and detailing these opportunities, based on the innovation architecture as a first step. In a second stage, the elements should be systematically evaluated on the strategic level for the formulation, in a third step, of an innovation strategy. In doing so, the focus should be on implementing an innova-

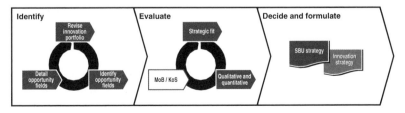

Figure 6.18 Project objectives of action research case at RubTec

tion strategy formulation process. This implementation should be actively supported by applying essential tools in the example process, in terms of 'learning by doing'.

The innovation architecture

The innovation architecture of RubTec was done on two different levels of detail. The more general innovation architecture, which is presented in Figure 6.19, shows an overview of the activities fields of the company. Twelve different strategic business fields could be identified that are based on nine strategic technology fields linked by functions. In the innovation architecture, three different cascades are integrated. The first cascade, technology intelligence, is based on methodological knowledge that allows the making of patent analysis and joint venture negotiation. This cascade is therefore a highly important element in the innovation system of the company, because it is a strategic decision that most of the technologies required to develop an innovation are bought externally. The next cascade is responsible for transforming the strategic technologies into specific products in one of the 12 strategic business fields. Based on these products, a marketing concept is developed on the next cascade to launch the products on different markets.

In contrast to the more detailed innovation architecture, indicated in Figure 6.20 for the technology side, the strategic technology platforms are detailed by specific technologies. These technologies vary in terms of products, material, design, evaluation or process technology. On the market side, the specific actual and future products are identified and integrated into the different strategic business fields. Because of the fact that the detailed innovation architecture is very large, segmentation has been done by the functions. Therefore, an innovation architecture was developed for each function.

The innovation strategy formulation process

Identify

During the process of architecting in the step 'identify', it was no problem to integrate the actual products and technologies as well as the future technologies. But, at the same time, it was obvious that in the traditional strategic business fields and strategic technology platforms only highly incremental innovations were planned. In these domains, no innovation with a medium or high degree of newness could be identified. In contrast, in the new business fields and technology platforms, where the first products are still in the phase of devel-

175

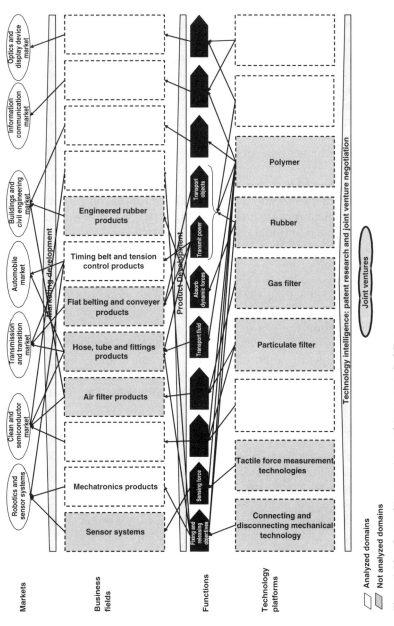

Figure 6.19 General innovation architecture at RubTec

opment, the innovations seem to have a high degree of newness. Therefore, the company has to manage two different kinds of innovation opportunities. On the one hand developing highly incremental innovations for business units that create already an important volume of sales but with a low margin, and on the other hand developing innovation with a high degree of newness in business fields that create today no, or a minimum, volume of sales, but with a high potential for the future.

After the innovation architecture was developed, different innovation fields were defined through a creative brainstorming, to identify new potential business fields. An example of the identification of new business fields is shown in Figure 6.20. This innovation field of 'pressure detection', derived from the business fields, is of 'sensor systems'. During the process of identification, the rule was defined that only ideas in this innovation field that are linked to actual strategic business fields were accepted. This rule would encourage the search to take place in traditional business fields that are today marked by incremental innovations, for the identification of potential innovation with a high degree of newness. The results were amazing. As presented in Figure 6.20 for the business field of belting, the idea came up to develop a belt that allows the position of a specific object on it to be detected that would allow for the elements on the belt to be retraced. These special ideas could be developed based on leveraging existing or planned technologies.

After the identification of potential strategic business fields, these innovation opportunities were detailed and integrated into the innovation architecture. It was not possible during the project to detail all the innovation opportunities therefore only some pilot projects were done.

Evaluate

For the evaluation of the innovation opportunities in the innovation architecture, firstly, a strategic fit was executed. The first-order fit had already been done in the step 'detail innovation opportunities', to ensure that the innovation architecture itself was consistent. The second-order fit was based on a core competencies check to identify whether all the innovation opportunities supported the core competencies of the whole company. For example, one idea was to develop a force-sensitive blanket for testing the characteristics of mattresses. This idea could be developed based on existing technologies, therefore the first-order fit is ensured. But in terms of second-order fit, this innovation opportunity would require a new sales system to be constructed, because the actual sales system is very strongly oriented to the mechanical industry, which is one of the core

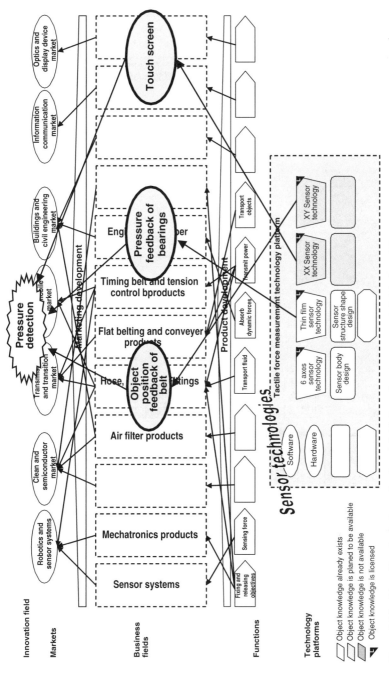

Figure 6.20 Identification of new business fields based on innovation fields at RubTec

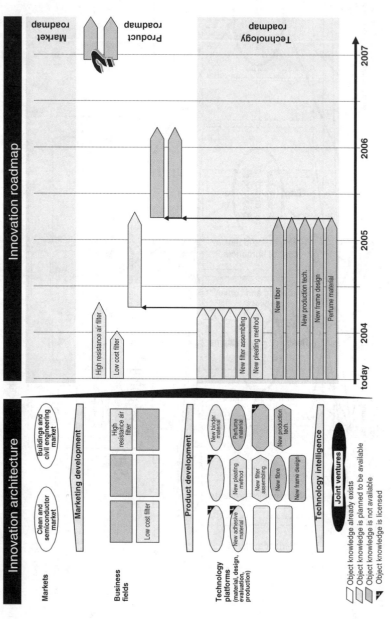

Figure 6.21 Innovation roadmap at RubTec

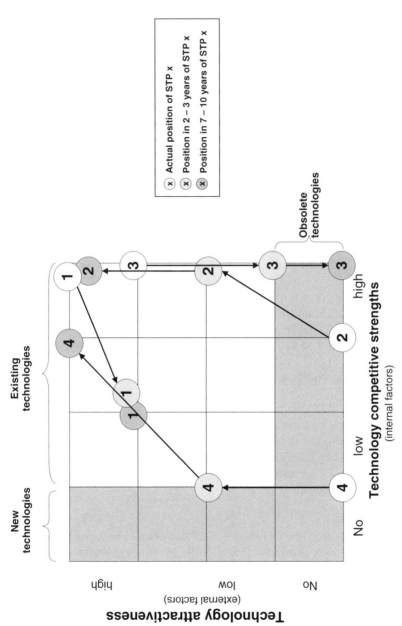

Figure 6.22 Technology portfolio at RubTec

competencies of RubTec. Therefore, the second-order fit is not ensured. The third-order fit was done by identifying the future trends of the industry and combining these trends with the innovation opportunities. Based on these trends, the employees could see that many innovation opportunities were leading in the right direction but, at the same time, some important trends were as yet being ignored.

After the strategic fit evaluation, the innovation opportunities were evaluated quantitatively and qualitatively. The innovation roadmap was elaborated for several functions of the innovation architecture in the manner of Figure 6.21. This development of the innovation roadmap was seen by the project members as very important, because such planning, in terms of an overview, had not been done explicitly in the past. This gave management the opportunity to see a realistic, integrated planning of innovation opportunities. Additionally, in the process of roadmapping, the employees have to communicate and align their activities more precisely. The innovation roadmap was completed by a knowledge gap analyzed to identify the importance of the planned objects.

The results of the evaluated innovation opportunities were partly summarized in the dynamic technology portfolio (see Figure 6.22). The results showed that, for example, as can be seen in Figure 6.22, the technologies 4 and 2, which are linked with quite new business fields, have a high potential in the future. But, at the same time, technology 3, which is a technology linked to a business field that is highly successful at the moment, will in 7 to 10 years be dramatically less attractive. With this technology portfolio, a first sense of urgency could be created.

Decide and formulate

The 'decide and formulate' step could not be undertaken completely because of the fact that the innovation architecture and the evaluation was not executed for all the activity fields of RubTec. Therefore, it was only possible to create a sense of urgency to close the gaps in the 'identify' and 'evaluate' steps.

Conclusion

Concluding, this action research case was one of the most important cases, because it showed – although the identification and evaluation could not be done for the whole company – that the innovation strategy development process, based on the innovation architecture, is a practical and structured solution for companies. This was clearly approved by statements of the company management which has accepted the implementation of this process of formulating an innovation strategy.

Additionally, this case indicated that this concept is not only a solution for formulating an innovation strategy in the culture of the western hemisphere, but also in Asian companies and, in particular, in Japanese companies.

Case 8: MicroSys

Short introduction

MicroSys is an independent company with 4000 employees with several business units. Its products are precision optical solutions based on microscopes and related instruments. MicroSys manufactures a comprehensive portfolio of products used in a wide variety of applications requiring measurement, analysis or lithography, the material sciences, industrial inspection and the semiconductor manufacturing industry.

The action research case was undertaken in the business unit, active in the medical industry. The products are microscopes for the surgical and diagnostic applications, sold throughout the world. The revenue is about 100 Mn, with a continuous annual growth of 3 per cent. The innovation budget is 4 per cent of revenue which is, compared to the three main competitors, below average. This low innovation budget is a concern for R&D, because the future success will mainly depend on the innovation ability for developing new products beside an active sales channel.

Initial position

Strategic innovation structures: The analyzed business unit of MicroSys has a process oriented organizational structure. This allows management to have a very good overview of the several activities in the business unit. Especially, it must mentioned that, in the innovation system, the structures for conducting professional technology intelligence and product development are effective and efficiently designed. Nevertheless, the structure of the innovation system has, according to management, the weakness that an explicit technology development is missing. This causes the problem that vague technology issues, discovered by technology intelligence, are not analyzed in detail in a separate process. Therefore, it often happens that the issues are no longer considered, although they may be of increasing importance.

Strategic innovation behaviours: MicroSys is known in the market as one of the companies with very high quality awareness. The developed products often have an unsurpassed level of quality. In addition, the development of the products is driven by a very strong market oriented culture. Therefore, the development of new products is mainly driven by actual customer needs rather than by new technology performances. Although the information flow on the market side is very enlarged, the communication between the technology and the market sides is not optimal.

Strategic innovation objectives: The strategic innovation objectives are, in short and middle-term, to enlarge the innovation ability in the existing activity domains. In the long-term the objective is to enter into new markets that are related to the actual activities. To fulfil the short- and middle-term objectives, a clear product strategy is defined. However, a technology strategy or an over-spanning innovation strategy, containing technology as well as product aspects, is missing.

Innovation decision processes: As mentioned before, the specific business unit of MicroSys is very market oriented. Therefore, the decision processes in innovation are also highly market oriented. New potential technologies are often not considered and a clear technology evaluation process is missing.

Project objectives: Due to the fact that at MicroSys the innovation decision processes for technologies is not highly developed, this process is updated especially with technology specific tools. Therefore, firstly, the ideas of technology intelligence are more structured by using the opportunity landscape. Secondly, the innovation opportunities are revised and detailed in the innovation architecture. Thirdly, it was decided to conduct a strategic fit evaluation in terms of Porter's three strategic order fits. This would show marketing the importance of certain technological issues and improve the overall innovation decision process.

Strategic intelligence

Strategic intelligence is completed by the opportunity landscape.[12] The opportunity landscape is divided into four strategic sectors: 'medical functionalities', 'optic and image property', 'handling and steering'

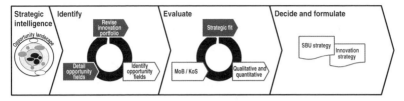

Figure 6.23 Project objectives of action research case at MicroSys

and 'construction and additional functionalities'. In a workshop, the opportunity landscape was filled with 26 issues in the various strategic sectors. The feedback from the workshop members was positive, because it allowed a differentiated visualization of all the issues, which were mainly technology driven. In their mind, this visualization would help to show marketing the importance of several technology based issues. This opportunity landscape was used as basis for the innovation architecture, which is the subject of the next section.

The innovation architecture

The innovation architecture of MicroSys, presented in Figure 6.24, has four different cascades. The competence development cascade is basically driven by the monitoring of scientific research fields for identifying new technologies and developing competencies in these technologies. Based on these identified technologies, regrouped in nine technology platforms, the modules are developed in the next cascade. The methodological knowledge of this cascade consists mainly of construction, patent creation and creativity methods. The modules are integrated on the next cascade, product development. This product development needs methods to define interfaces, analyze the user characteristics, create ergonomic designs and ensure quality certification. At the top cascade the products are introduced into the market by marketing.

In this innovation architecture, the decision was taken to integrate a two level detailed visualization of the functions. The more general level consists of four functions, which are the main product functions. This more general definition allows the strategic direction to be defined in terms of future innovation activities on a more abstract level. But due to the fact that these four functions are too general for steering specific innovation projects, the decision was taken to detail the general functions into 17 detailed functions.

184 *Structured Creativity*

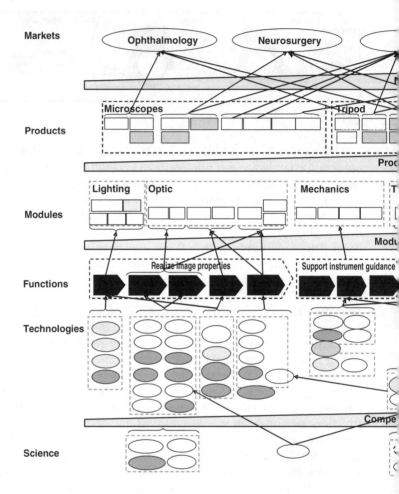

Figure 6.24 Innovation architecture at MicroSys

The innovation strategy formulation process
Identify
The 'identify' phase mainly consisted of revising the innovation portfolio based on the issues integrated into the opportunity landscape and detailing these new issues in the innovation architecture. During this process, interviews were conducted with employees from marketing, product management and R&D. A first version of the innovation architecture attempted to realize the consistency of the different innovation

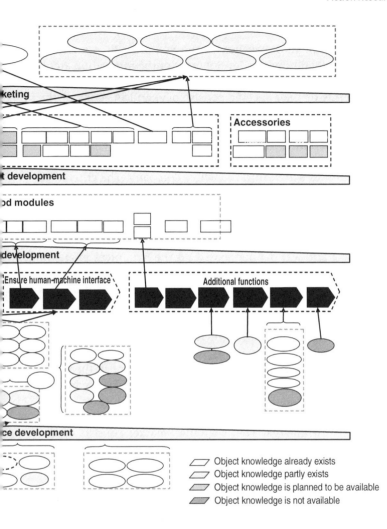

opportunities from the market side, as well as from technology side. In a second version of the innovation architecture, the functions were defined. The technology oriented project members mentioned that this procedure helped them to visualize their ideas in context. Therefore, it was possible for them to show the link between the technology and what it can do specifically in terms of a detailed function. They concluded that it is a good communication tool to bring technological issues better into the product development process.

Evaluate

The strategic fit evaluation began with the first-order fit, which was already done mainly during the development of the innovation architecture. Nearly all the objects in the innovation architectures could be analyzed in terms of their specific consistency, and in terms of the functional handshake. Only some technologies that were fulfilling additional functions could not be integrated completely into the innovation architecture during the project, as can be seen in Figure 6.24.

The second-order fit was done by evaluating how the different innovation opportunities will fit into the whole company. The criteria for this evaluation were based on technological fit, organizational fit, strategic fit, and financial fit.

The third-order fit was done with the scenario technique. The future scenarios were realized from two different points of view. Firstly, a competition scenario was defined, based on the five forces model of Porter (1980: 26). Each force[13] was analyzed in a creative workshop, studying how the future could change. For example, in the future it is possible that a substitute competitor will appear in the pharmaceutical industry because they reduce the diseases without making operations, where MicroSys microscopes are used. The second point of view was taken from a marketing standpoint. Here, the changes in the markets were analyzed. According to the workshop members, one of the major changes, for example, would be that the market will be divided into a 'forever young' market and a 'spartanic' market. The 'forever young' market consists mainly of hospitals doing plastic surgery conducted with high tech instruments. The 'spartanic' market consists of the public hospitals that are under cost pressure as a result of health reforms. The cost factor for the surgical instruments is the most important element in this market. Based on the two different scenarios, a preliminary discussion took place, comparing the scenarios to the innovation opportunities in the innovation architecture.

According to management, the strategic fit evaluation allowed, firstly, the technologies that were often in the past decoupled from the company context to be more precisely integrated. Secondly, due to the scenario technique, the innovation opportunities could be evaluated in the long term. This is a major advantage for communicating with marketing as it more precisely shows the importance of a technology. Thirdly, it was noticed that the actual strategies of MicroSys do not respond entirely to the scenarios that were identified as important in the future.

Conclusion

A conclusion of this action research is that the innovation strategy development process can be perfectly coupled with strategic intelligence. Another major awareness was that scenario planning shows the importance of technologies in the long term. This tool is therefore especially important for companies that are highly market oriented, and therefore the evaluation of innovation opportunities is often based on short term criteria.

Innovation architecture as basic tool, as well as the linked opportunity landscape, and the scenario technique, were found highly attractive by management. Convinced by these tools, management decided to use innovation architecture more intensively in the future by also implementing the innovation roadmap.

Case 9: StockTec

Short introduction

StockTec is a multinational company that specializes in specific products for supply stock systems. With its products, it is the world leader, with a market share of approximate 50 per cent. In 2002, about 450 employees realized an EBIT of 3.6 per cent.

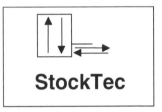

Stock Tec had three major challenges to master during the running of the action research case. Firstly, the complexity of the products is, compared to the competitors, very high. Secondly, the transparency of the activities is very low. And thirdly, the company processes are causing many problems. These problems were identified in the value providing system, as well as in the innovation system, as is now presented.

Initial position

Strategic innovation structures: The innovation system consists of an R&D department, without having a strategic marketing section. This lack of strategic marketing leads to the fact that sales directly provides the innovation system with short-term customer needs and, for medium- or long-term innovation opportunities, R&D itself decides what to do. Because of the structural lack, it happened that a new product was developed that was obviously not according to the customer needs.

Additionally, in the company no person was found that is responsible for deciding whether a new product is required or an old product can be taken out of the catalogue.

Furthermore, the structure of the innovation system does not easily allow a distinctive development of standardized and flexible customer specific products. The standard products are therefore often too expensive, because they are too flexible.

Strategic innovation behaviours: The employees in R&D are very technocratic. Technological ideas are often integrated into new products, without checking whether the customer actually requires these technologies. This technocratic behaviour allows the company to get an international logistic award for its product, but the customer did not buy the product as planned.

Strategic innovation objectives: The main strategic innovation objective is, first of all, to ensure in the future that a standard mass product can be developed that fulfils the requirements of most of the customers. Such a product should be obviously less expensive and less complex than today's products. But, at the same time, StockTech wants to ensure that also customer specific products can be developed in the future.

Innovation decision processes: The decision process for deciding about future innovation activities is not really in place. The fact of the matter is that StockTec has nearly no data to provide for such a decision-making process. Therefore, decisions are taken based on opinions.

Project objectives: Based on the above problems of StockTec, the main project objective is to define the required innovation processes and organization. This innovation organization should be in alignment with the strategic innovation objectives of today and should allow development of more market oriented products. To reach this project objective, a two step procedure is put in place: Firstly, the innovation architecture is con-

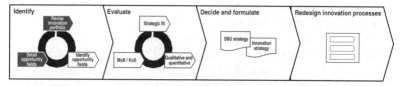

Figure 6.25 Project objectives of case at StockTec

structed by revising and detailing the innovation opportunities for deriving, in a second step, the innovation processes and their organization.

The innovation architecture

The innovation architecture of StockTec was firstly constructed in a detailed version, and then reduced to the essential elements for deriving the innovation processes. The reduced innovation architecture is presented on the left of Figure 6.26. This innovation architecture is marked by three cascades. The first cascade is the module development, whereas a two-folded distinction in terms of its methodological knowledge is done which consists, on the one hand, of knowledge to coordinate the external development of modules and, on the other hand, of knowledge to develop modules internally. Compared to other innovation architectures, at StockTec there is no technology development cascade, because the company does not have the financial resources. Therefore, new technologies are developed externally. The next cascade is product development, which is differentiated by mass product and customized product development methodological knowledge. This differentiation is based on the fact that the methodological background is different. For example, the methods for mass product development require an intensive consideration of the production costs. In contrast, the development of a customized product requires greater consideration of development costs. These are two different methodological mindsets that cannot be done in one and the same process. The top cascade is the marketing cascade for ensuring the short, medium- and long-term customer needs to be captured, and the market introduction to be prepared.

The innovation strategy development process

Identify

In the 'identify' phase, the innovation architecture was constructed. It was astonishing that, in a first draft of the innovation architecture, the future products were not differentiated between mass and customized products. R&D was even planning to develop highly flexible products that could be used for customized products as well as for mass products. This intention would have contributed to the effect that the mass products would have been too expensive and the customized products would have been too cheap. Innovation architecture was therefore designed so that it was in alignment with the strategic innovation objectives. The future products to develop were differentiated into two categories, in order to have a consistent innovation architecture.

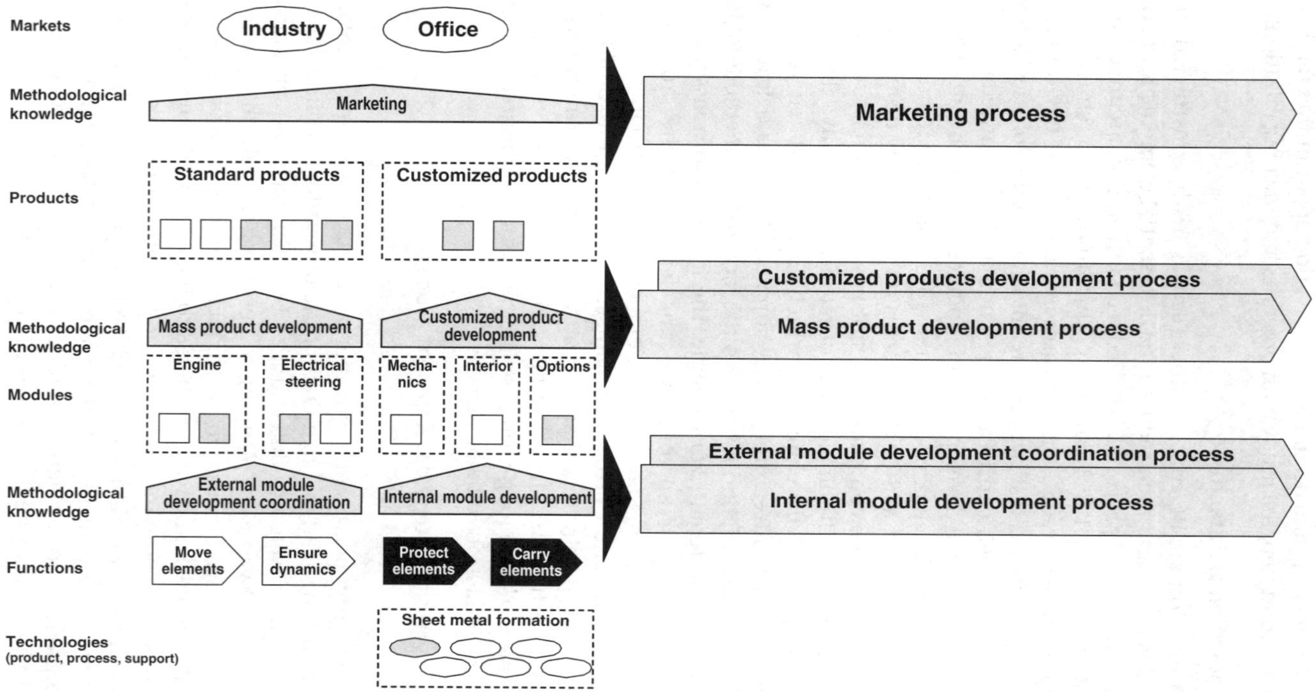

Figure 6.26 Deriving the innovation processes based on the innovation architecture at StockTec

Normally at this point, the different objects should be evaluated. But, due to the fact that management did not have enough data to be able to evaluate all the product ideas, it was decided as a first step to design the organization direct. Management was thereby informed by the fact that, for example, the decision to not make any more customized products would also influence the organization.

Redesign innovation processes

Based on the innovation architecture of StockTec, which was not entirely decided on a strategic basis, the innovation processes were derived. As shown on pp. 121ff. the innovation processes are derived based on the cascading and segmentation of the methodological knowledge, as shown in Figure 6.26.

These innovation processes already define the organizational units. Therefore a 'marketing', a 'customized product development', a 'mass product development', an 'external module development' and an 'external module development' organizational unit were planned.

The innovation processes of Figure 6.26 needed to be detailed in a further step in terms of their interaction and process steps, as shown in Figure 6.27. The interactions between the processes are always based on an order–deliver understanding. For example, the marketing process decides to make a new mass product and gives this order to the mass product development process, which has to deliver the defined product. This mass product development process then decides to integrate a specific module, and gives an order to the internal module development process in the event that they can do it; otherwise, the order goes to the external module development process. All of the processes are steered by a strategic innovation management process which, in fact, consists basically of the strategic process phases of Figure 5.14.

Conclusion

The action research case at StockTec showed explicitly that innovation architecture is not only a tool that allows strategic innovation opportunities to be identified in a structured manner, but also the organizational structure can be defined. Thereby the organizational structure is not defined based on a theoretical organization structure or a best case organization in the industry, but is based on the company specific activities. This is a major advantage and helps to ensure the efficiency and effectiveness of a company.

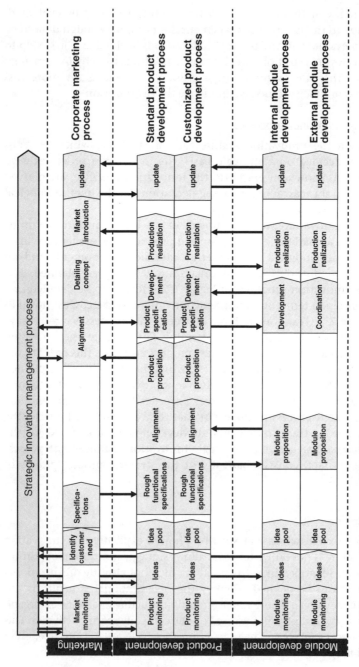

Figure 6.27 Innovation processes at StockTec

Cross-case analysis and conclusion

Cross-case analysis of the innovation architecture

The different innovation architectures of the action research cases show very well that for small and large companies, as well as for turnaround and successful companies, an innovation architecture could be constructed according to the theory presented on pp. 57–87.

The **concept** of innovation architecture ensures that the structure of the architecture is aligned to the specific activities of the company. For example, the companies that are active in research have more cascades than companies that do not. Also, the segmentation of methodological knowledge is adapted to the analyzed company. Both, object as well as methodological knowledge, was very important in all the cases in order to understand the structure of the innovation architecture. In contrast, the meta-knowledge, which completes the image of knowledge in the innovation system, was in general not seen as so important in constructing up the innovation architecture. The reason for that is perhaps due to the fact that meta-knowledge is today often not considered in strategic decisions.

The **visualization** of the innovation architecture was basically the same, but varies in details from case to case. This is because the visualization of the innovation architecture had to be adapted to the specific requirements of each project. Therefore, based on the specific projects, some innovation architectures were built up on one page, others were segmented by products or functions, and others were more detailed or more general. This shows that innovation architecture has, on the one hand, very strict rules for construction but, on the other hand, there is enough flexibility to adapt it to the specific project need.

The **usability** of innovation architecture was, in general, considered in the action research cases as very high. The first creation of innovation architecture is a step which is associated with an intensive work phase, which decreases according to how well the company is already structured. However, after the innovation architecture is constructed, the update can be easily achieved, as proven in one case. Also, one case showed that management can easily develop the innovation architecture for itself. Therefore, in all the cases innovation architecture seems to be a very practical tool.

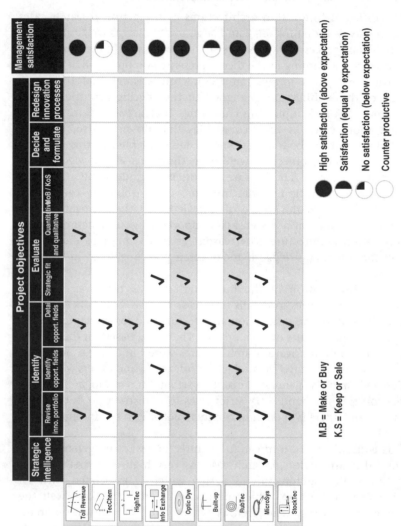

Figure 6.28 Cross-case overview of project objectives and management satisfaction

Action Research 195

The **results** of innovation architecture seem to be, in the context of innovation management, highly multifaceted, as presented in the following:

- Innovation architecture is a basic tool for structuring the innovation systems of companies that are highly unstructured, as is often the case in turn-around companies. At the same time, innovation architecture helps well-structured companies to visualize the actual system, as well as the possible future system, by systematically integrating the potential innovation opportunities.
- Innovation architecture is a basic tool for aligning market pull activities, mainly initiated by marketing, and technology push activities, often initiated by research and development. Thereby the functional handshake between products/modules and technology is a major element. But also the fit between all the objects is important, to ensure the alignment of all the intentions and activities in different domains of the company.
- Innovation architecture is a basic strategic steering tool in innovation management. Firstly, due to the fact that the knowledge gaps are illustrated, it is clearly shown that they have to be closed. Secondly, the functions allow a solution neutral responsibility for the domains of technology development to be defined. Thirdly, innovation architecture supports strategic steering to develop highly systematic and purposeful innovations from scientific science to the market introduction.
- Innovation architecture is a basic tool that integrates all innovation opportunities independent from its degree of newness or object.[14] This allows, in one visualization, a holistic overview of the whole innovation system.
- Innovation architecture is a basic tool for enforcing structured communication. Based on the innovation architecture a discussion, including market and technology oriented members, can be conducted in a more structured manner because everyone talks about the same elements while the innovation architecture is completely constructed.

To conclude, innovation architecture is a basic tool for innovation management that was confirmed by most of the partners of the action research cases.

Cross-case analysis of the innovation strategy formulation process

The innovation strategy formulation process was accepted by management in most cases, as presented in Figure 6.28. Seven of the nine action research projects were concluded very satisfactorily, which means that the results were above the expectations of management. One project ended with average satisfaction and one with no satisfaction. In both of these action research cases, management mainly criticized that the effort was greater than the results. Comparing the two cases with the other cases, these two cases were the only projects where the development of the innovation architecture was done solely to revise the innovation portfolio and to detail the innovation opportunities. In all the other cases, innovation architecture was used for further purposes, which allowed more results to be gained. This leads to a first awareness, that the satisfaction factor for innovation architecture tends to be higher when it is used more broadly in innovation strategy development process.

To complete the cross-case evaluation of the nine cases, the innovation strategy formulation process is presented in the following by discussing the concept, the usability and the results.

The **concept** of the innovation strategy formulation process ensures that management is directed through a procedure that allows all potential innovation opportunities to be identified and analyzed systematically and practically. Thereby the process itself seems to be more important than the results. Therefore, all decision makers have to be involved. But it must mentioned, as some cases showed, that before beginning the process of innovation strategy formulation, a general corporate or business unit strategic direction has to be present. This allows the evaluation criteria to be determined in the evaluation step, especially in the strategic fit evaluation.

The **usability** of the innovation strategy formulation process was strongly confirmed by practice. The cases showed that the process could be used as a modular basic concept that completes the missing steps in a company's innovation strategy formulation process, as well as a complete strategy formulation concept. Also, the cultural background of the company is not especially important, because the process was accepted in Europe as well as in Japan. This positive feedback from management concerning the usability is probably based on two facts. Firstly, the selection of the tools, consisting of innovation architecture and general management tools, helped in considering all the important aspects for formulating an innovation strategy. Secondly, many tools integrated into the process were already known by management, which allowed the tools to be implemented much faster.

The **results** of the innovation strategy formulation process seem to be highly multifaceted in the context of innovation management, as is presented in the following:

- The innovation strategy formulation process allows a clear strategic direction to be defined for the innovation system, consisting of objectives and a path, without reducing the creativity of the individual person. This is because innovation architecture allows evaluation to be made on a level that is detailed enough for evaluation, but not so detailed that the complete product is already entirely designed.
- The innovation strategy formulation process ensures many gaps are identified before starting the development of an innovation opportunity. These gaps are, on the one hand, knowledge gaps, time gaps and resource gaps, and, on the other hand, gaps in terms of missing links between objects can be easily found, as management often pointed out.
- The innovation strategy formulation process is not only a process of decision making, but it is also a process that initiates the planning and implementation phases of specific innovation opportunities. For example, the innovation roadmap is a tool that helps to prepare a decision, for identifying gaps in terms of timing, but also this innovation roadmap is a detailed input for planning and implementation. Therefore, this innovation strategy formulation process does reduce the gap between strategic decision making and operational implementation.
- The innovation strategy formulation process allows, as seen in one case presented, the deriving of appropriated organizational processes in the innovation system. This ensures the efficiency of the organization in the innovation system.
- The innovation strategy formulation process ensures, on strategic level, the required creativity to be able to identify all potential innovation opportunities but, at the same time, the evaluation is marked essentially by a fact based reduction of the innovation opportunities.

Finally, the innovation strategy formulation process was seen by a majority of management as a process that is very useful. Therefore, the call from reality presented in Chapter 3 can be answered:

> The innovation strategy formulation process developed in this work is a solution for a structured practitioner-oriented innovation strategy formulation process for innovation driven companies.

7
Conclusion and Working Hypothesis Rethought

In the previous chapter, it was presented that innovation architecture and the innovation strategy formulation process is highly accepted by management practice. However, it has still to be evaluated whether the innovation strategy formulation process is in alignment with the theoretical requirements of Chapter 4. This evaluation of the concept is shown in Figure 7.1 and described in the following.

The innovation strategy formulation process allows a **strategic management specific understanding** of complexity and systemic interaction to be provided, because the innovation architecture integrates object, methodological and meta-knowledge in a visualization model, which ensures a reduction of the complexity and a visualization of the systemic interaction. The evolution is understood because the potential innovation opportunities are integrated into the innovation architecture and compared to the environmental trends.

Combining these theoretical insights and the above presented results of the action research cases, the **first working hypothesis** presented in Chapter 4 can be validated at this point:

> The concept of architecture is a solution for understanding the complex, systemic interactive and evolutionary system of innovation driven enterprises.

The innovation strategy formulation process allows provision of a **strategy specific understanding** of direction because the resulting innovation opportunities are analyzed in terms of resources, time, knowledge gap and make or buy/keep or sell, which leads to a clear strategic objective and path. The focus is ensured due to the fact that,

		Quinn 1985	Kawai 1992	Tschirky 1998	Afuah 1998	Martensen and Dahlgaard 1999	Innovation strategy formulation process
Provide a strategic management specific understanding of	Understand complexity	◐	◐	◐	◐	◐	●
	Understand systemic interaction	◐	◐	◐	◐	◐	●
	Understand evolution	◐	◐	●	◐	◐	●
Provide a strategy specific understanding of	Direction	●	◐	●	●	●	●
	Focus	◐	◐	●	●	●	●
	Organization	◐	○	○	●	◐	●
	Consistency	●	◐	●	●	●	●
Providing an innovation specific understanding of	Integral innovation	○	○	◐	●	○	●
	Innovation barriers	○	○	◐	◐	◐	◐
	Innovation newness	○	○	◐	○	○	◐
	Innovation relevant knowledge	○	○	○	●	◐	●

Legend: ● Explicit supported; ◐ Implicit or partly supported; ○ No support

Figure 7.1 Evaluation of the innovation strategy formulation process compared to the existing concepts

firstly, the resources are considered and, secondly, the innovation opportunities are aligned to the core competencies of the company. Based on the innovation strategy, the organization can also be derived easily by using innovation architecture, as shown in the case of StockTec. In addition, the consistency of the innovation strategy is ensured due to the strategic fit evaluation based on the innovation architecture.

The innovation strategy formulation process allows an **innovation specific understanding** of integral innovations to be provided, because technological and business innovations are integrated into the innovation architecture and, therefore, form part of the innovation strategy formulation process. The organizational innovations can be derived based on the innovation strategy, in a next phase. The innovation barriers that occur due to a lack of strategic fit, resources, time or knowledge are considered. But innovation barriers that are generated due to cultural or personal aspects, such as a lack of motivation, are not explicitly analyzed. At this point, the concept does not provide a complete solution. Innovations with a high degree of novelty and a low degree of novelty are, in general, considered in innovation architecture. However, some special potential innovation opportunities where, for example, it is known that a technology such as nano technology will be important for the industry, but a specific product is still not known, are too soon rejected. This is because a major condition in the evaluation process is that all the innovation opportunities must be completely detailed, and therefore the nano technology will not be considered in that innovation strategy. In such cases, the company has to pay attention. And, last but not least, the innovation-relevant knowledge, as basis for all innovation, is completely understood by innovation architecture and the knowledge gap analysis.

Based on the fact that the actual theory cannot provide a better solution in terms of explicitly considering all innovation barriers, and especially innovations with a high degree of novelty (see Figure 7.1), it can be said that this innovation strategy formulation process is, in general, an appropriate tool that respects the theory and is a practical and structured solution for practice. In a sum, the innovation strategy formulation process, which is based on innovation architecture, is also an appropriate solution for companies from theoretical point of view and, therefore, the **second working hypothesis** presented in Chapter 4 can be validated at this point:

> The innovation architecture applied in an adapted innovation strategy formulation process is a support for innovation driven companies enabling them to define an innovation strategy.

However the innovation strategy formulation process needs to be implemented proficiently into the company specific conditions. In this context, the **third working hypothesis** presented in Chapter 4 can be validated, based on insights gained in the action research cases:

> Implementing an innovation strategy formulation process in an innovation driven company is not realizable by implement-

ing the whole theoretical process. However, the theoretical innovation strategy formulation process is a basis for adding the missing steps, so that the company will be able to define an adequate innovation strategy.

In a nutshell, the innovation strategy process based on innovation architecture is, from a theoretical as well as from a practical point of view, a practitioner-oriented and structured solution for closing the dual gaps presented in Chapter 4.

8
Towards a New Set of Management Principles

The aim of this chapter is to make a major contribution towards fulfiling the need in practice (see pp. 60ff.). This need represents a call from reality for management guidelines, which could support the practitioner in designing and implementing a practitioner-oriented and structured innovation strategy formulation process in innovation driven enterprises. Therefore, this chapter is the practical answer to the three research questions on how to understand a complex, systemic interacting and evolutionary innovation systems, and how to design and implement a practitioner-oriented and structured innovation strategy formulation process.

The nature of this chapter is different from the character of the other chapters. While the latter follow scientific and sound argumentation, output in this chapter is normative and hands-on. Therefore, proposals in this chapter are not thoroughly argued, but quite straightforward. Any suggestions are based on practical experience gained during this research and a sound theoretical background in the field of innovation strategy formulation. This also implies that the principles do not strictly depict the elaborated solutions offered during the action research case studies. This would be contrary to the most important insight of recent research and this study; namely, a strong dependency between the structure of the innovation strategy formulation process and the company's context. Since the context differs from one enterprise to another, the management principles cannot be seen as recommendations on a very detailed level. Therefore, the management principles give general indications.

The following management guidelines seem to be promising for innovation strategy formulation in innovation driven enterprises. They are listed in an order that corresponds logically to the concerns of innovation driven enterprises when setting up an innovation strategy formulation concept.

The first three principles are basic statements to understand the strategic level of an innovation system. Then four clusters build a cluster of principles to formulate an innovation strategy for innovation driven companies. Furthermore, a cluster with three principles makes suggestions for the implementation of an innovation strategy formulation process. This discussion should be understood from the point of view of an innovation driven company. Therefore, key benefits and arguments are related to innovation driven contexts. However, some principles or parts of principles would be of interest to any company. To underpin the ten management principles, each principle underlies a citation of a known scholar in the domain of general management. Therefore, the sample of these management principles in the context of innovation management is a main finding of this work.

Understanding the innovation system

Principle 1: Design an architectural blueprint

> Innovation driven enterprises should understand their innovation system in terms of complexity, systemic interaction and evolution. For this purpose, innovation architecture is an appropriate tool.

Key benefits:
- Management of innovation driven companies are able to understand their system, and therefore they can improve the quality of their decision-making
- Innovation opportunities can be analyzed more in detail
- A concrete plan for the future development of the company can be developed.

The concept of architecture is always a blueprint of a system which aims to understand it. Especially in the case of complex, systemic interactive and evolutionary systems, which cannot be understood by one individual, its essential elements can be better understood by developing an architectural blueprint. Therefore, the first step in the strategic decision process should be to develop an architecture.

Supporting citation of general management scholar:
'An architect must be capable of dreaming of things not yet created – a cathedral where there is now only a dusty plain, or an elegant span across a chasm that hasn't yet been crossed. But an architect must also be capable of producing a blueprint for how to turn the dream into reality' (Hamel and Prahalad, 1994: 107).

Principle 2: System analyzed with appropriate effort invested

For understanding an innovation system in terms of taking a strategic decision, it is more important to consider the innovation opportunities' evolution and interaction with others than the properties of the opportunity itself (e.g. functioning of the technology).

Key benefits:

- The innovation strategy formulation process can be designed effectively and efficiently
- Communication on a strategic level is based on the important elements.

To understand an innovation system, as well as a system in general, it is not essential to comprehend all details. It is often not even necessary to understand the functioning of each element of the system. Nonetheless, it is important to understand the interaction of an element with others, and its evolution. In the case of formulating an innovation strategy, it is not important to know the functioning of a specific technology, but it is important to know its potential usability, its costs, or its future attractiveness and so on. Additionally, for a corporate innovation strategy, the break-up of a technology platform into specific technologies need not to be as detailed as for the innovation strategy of a smaller business unit, which results from the different requirements to a strategy.

Supporting citation of general management scholar:
'The central concept "system" embodies the idea of a set of elements connected together which form a whole, this showing properties which are properties of the whole, rather than properties of its component parts' (Checkland, 1993: 3).

Principle 3: Think in functions

> For developing an innovation strategy it is essential to know the product functions.

Key benefits:
- Functions allow a solution neutral alignment between market and technology oriented activities in terms of taking decisions as well as in communication
- Functions allow responsibilities to be defined without decreasing the creativity for finding solutions
- Functions allow new business fields or technology platforms to be identified based on existing activities
- Functions, especially core functions, very clearly define the future direction of the innovation system.

In innovation architecture, the functions make the solution neutral link between market pull and technology push. The management of the action research cases found this to be an important link because, for several reasons, its definition helps to steer an innovation system. Firstly, the communication between marketing, research and development can be enforced through a solution neutral but very specific discussion of developing an innovation. Secondly, functions allow a solution neutral identification of possible future opportunity fields. Thirdly, functions allow new potential business fields as well as new potential technology platforms to be identified. Fourthly, the strategic definitions of core functions allow the company to direct the innovation system to a specific direction without repressing creativity.

Supporting citation of general management scholar:
'In reality a product should be considered simply as a physical manifestation of the application of a particular technology to the satisfaction of a particular function for a particular customer group' (Abell, 1980: 113).

Formulating an innovation strategy

Principle 4: Ensure strategic fit

> The strategic fit evaluation of innovation activities is essential to develop innovations based on actual competencies.

Key benefit:
- Innovations that are based on actual competencies have, in general, a higher probability of succeeding with less effort.

An innovation strategy that defines future innovation activities that are not supporting the companies actual activities or the environmental trends are, on the one hand, not supporting the core competencies of the company which allow competitive advantage and, on the other hand, they need much more effort to be introduced into the market successfully. Nevertheless, it must be mentioned that innovation opportunities without strategic fit can also be successful. However, in such a case management must be clearly aware of the consequences of such a decision. Therefore, in general, innovation opportunities with a clear strategic fit are more interesting.

Supporting citation of general management scholar:
'Strategy is creating fit among a company's activities. The success of a strategy depends on doing many things well – not just a few – and integrating among them. If there is no fit among activities, there is no distinctive strategy and little sustainability' (Porter, 1996: 75).

Principle 5: Direct and focus with consistency

An innovation driven enterprise should give its innovation system a direction in terms of a specific objective and a predefined path. This direction should allow activities to be focused and consistency to be ensured.

Key benefits:
- Innovation driven companies can ensure their implementation of their strategic intentions
- Duplications or inconsistencies are reduced to a minimum, and this ensures an effective and efficient realization of innovations.

A strategy that is unable to be communicated to every employee in the company will not be a lived strategy. Furthermore, a lack in direction of the strategy will not allow the efforts to be focused on the major goal. This will result in inconsistency in the implementation or increase the inconsistency in the case where the strategy is not consistent. Therefore, only when a strategy defines direction, in terms of objectives and path, ensures focus and is consistent will a strategy be a lived part of the company.

Supporting citation of general management scholar:

'A strategy is a plan or a pattern that integrates an organization's major goals, policies, and action sequences into a cohesive whole' (Quinn (1980: 7).

Principle 6: Define clear responsibilities

An innovation strategy of an innovation driven enterprise should point out clear responsibilities of identifying, evaluating and realizing opportunities.

Key benefits:

- Tasks are clearly allocated which ensure their efficiency
- The controlling of the implementation of an innovation strategy can be simplified.

Defining clear responsibility in strategies is an essential element to ensure its implementation because of several effects. Firstly, a strategy that defines no responsibilities will leave a major gap between strategic level and operational level, because the following question will be raised but not be answered: Who will do this? Secondly, project responsibilities that are clearly defined ensure that a clear input is defined and, more importantly, a clear output is defined, which focuses the work on realizing and not redefining. Thirdly, a clear responsibility for tasks establishes control of the task.

Supporting citation of general management scholar:

'Responsibilities for accomplishing key tasks and making decisions must be assigned to individuals or groups' (Andrews, 1987: 87).

Principle 7: Optimize the organization

> The organization of the innovation system should be adapted to the innovation activities.

Key benefits:

- An organization that fits to its activities has a high level of effectiveness and efficiency
- An organizational redesign in terms of an organizational innovation ensures, in the context of integral innovations, sustainable development and therefore competitive advantage.

An innovation strategy describes what the company has to develop in the future. Therefore, it is obvious that a changed strategy will influence the organization to some degree. For example, the decision of a company specialized in mass production to develop products for special customer needs will influence the innovation organization, because it is not possible to develop a mass product with a focus on production costs and a specific product with a focus on high flexibility in the same department. Therefore, it is essential to design the company's processes according to the strategic intentions, and to design the structure of the organization according to the processes.

Supporting citation of general management scholar:
'Structure follows Strategy' (Chandler, 1962).

Implementing an innovation strategy formulation process

Principle 8: Define continuous process ownership

A process ownership of the innovation strategy formulation process is essential for its successful implementation.

Key benefit:

- With ownership of a process, responsibility for its outcome and its practical use is enforced.

The definition of a continuous process ownership has, in the context of this work, a two-fold meaning: firstly, for the innovation strategy formulation process a member of the company board should take the ownership and the detailed output of this process has to be defined. Thus, the innovation strategy is formulated or updated periodically and, in the same structure, the innovation architecture is updated as well. Secondly, a continuous process ownership is also essential for all the processes in the innovation system. For example, one person should own the technology development process, and this person should therefore be responsible for every aspect from the intelligence to the development of a technology. This defines a clear responsibility and supports creativity in the process.

Supporting citation of general management scholar:
'Through a strict design of continuous processes in the company, through clearly defined process outputs and through a clear regulation of the responsibilities it is possible to gain ... major benefits of economizing' (Horsch, 2003: 14).[1]

Principle 9: Nurture participation

> The integration of ideas from marketing and sales, development and research, and production helps to expand the horizon for innovation strategies and facilitates its implementation.

Key benefit:
- Ideas from marketing and sales, development and research, and production help to provide a more holistic understanding for future innovation activities
- When an innovation strategy includes internal ideas, the employees' identification with the innovation strategy is stronger.

In a company, many ideas are already internally available to prepare the formulation of an innovation strategy. Therefore, it would be a pity to ignore all these ideas. However, it is often a problem for management to have the time to hear all of the often unstructured ideas. Nevertheless, there are tools, such as the opportunity landscape and innovation architecture, which nurture participation on an operational level by summarizing the ideas. Afterwards, these ideas can be evaluated on a strategic level and, if it is necessary, with further support from the operational level. This kind of participation allows new strategic directions to be found internally. Furthermore, an innovation strategy that is based on internal ideas is supported by the company's own employees, which helps to implement its adoption.

Supporting citation of general management scholar:
'On complex projects, the inner team cannot sustain itself and work effectively without constantly importing new information from the outside world' (Allen, 1977: 122).

Principle 10: Develop a culture of discipline

> A strong management commitment to a clear and focused innovation strategy is essential to ensure the innovation strategy formulation process and the implementation of the strategy itself.

Key benefit:
- A strong management commitment ensures a focus of the activities, which makes them more effective and efficient.

In the event that all the above-mentioned management principles for the innovation system were followed, one highly important principle is still missing: the company has to have the culture of discipline to implement strictly, with every effort, exactly what they have decided. If this is not the case, the innovation strategy formulation process, as well as the innovation strategy, will not be implemented. Therefore, the presented innovation strategy development process based on innovation architecture will only be successful – as is the case for all management tools – if the company implements it by focusing on that which is essential for fulfiling the strategic goals.

Supporting citation of general management scholar:

'Everyone would like to be the best, but most organizations lack the discipline to figure out with egoless clarity what they can be the best at and the will to do whatever it takes to turn that potential into reality. They lack the discipline to rinse their cottage cheese' (Collins, 2001: 128).

Notes

1 Introduction

1 Some exemplary authors: Schumpeter (1927), Zahn and Weidler (1995: 359), Schaad (2001: 1), Bucher (2003: 2).
2 Effectiveness is a qualitative factor and is affected by culture, capital, organization, environment, quality of education and experience, science, technologies, knowledge base, quality of information and through meta skills (Zahn, 1995: 189f.).
3 Efficiency is related to costs, needed to implement a specific degree of effectiveness in the market proposition (Zahn, 1995: 190).
4 The authors have translated the German word 'dispositiv' used by Hauschildt (1997: 25) with anticipative.
5 Innovation driven companies need successful commercialized renewals to gain competitive advantage and to be successful in the long term. Therefore, these companies are especially interesting when analyzing the subject of innovation.
6 Aregger (1976), Hauschildt (1997: 44ff), Afuah (1998: 99; 2002: 369), Tidd, Bessant and Pavitt (2001: 65), Bullinger and Auernhammer (2003: 29), Horsch (2003: 68). A more detailed discussion about innovation strategy can be found in chapter 2.
7 Yin talks about cases in general and not about action research cases. But the authors wish to point in this book that the cases are done in a specific manner, based on action research. Action research is an approach in which practitioners and scientists jointly design and implement new concepts. Moreover, the involved scientists try in turn to systematize and generalize their experiences (cp. Kubicek, 1975: 70).
8 Van Maanen (1979: 539). The author speaks about 'first-order and second-order findings'.
9 For organizational research in general, cf. Kubicek (1975), Kieser and Kubicek (1992) and Grochla (1978).

2 State of the Art in Theory

1 Compare also Teece (Teece, 1990: 40); Welge and Al-Laham (Welge and Al-Laham, 1992: 2356); Hauschildt (Hauschildt, 1997: 25); Maurer (Maurer, 2002: 17); Hunger and Wheelen (Hunger and Wheelen, 2002: 2).
2 A more detailed discussion about the term innovativeness can be found on page 18.
3 Translated from the German word 'Systemhaftigkeit' in Malik (2001a: 139); a concrete use of the term systemic interaction in the English literature can be found in Snyder et al. (1980) and Evangelisty et al. (2002).

4   For a more detailed discussion about evolutionary theories, see Hannan and Freeman (1977); Nelson and Winter (1982); Baum and Singh (1994); Pherson and Ranger-Moore (1994) and Barron (2003).
5   Compare for that purpose Marquis (1969: 1).
6   According to Sanchez (Sanchez, 2001), 'capabilities are repeatable patterns of action that an organization can use to get things done. Capabilities reside in groups of people in an organization who can work together to do things. Capabilities are thus a special kind of asset, because capabilities use or operate on other kinds of assets (like machines and the skills of individuals) in the process of getting things done'.
7   The following literature could be taken for a more detailed discussion about innovation barriers: Stahl and Eichen (2003: 16f.); Afuah (1998: 97ff., 217ff.); Biermann (1997: 38ff.); Bullinger and Auernhammer (2003: 34).
8   Gabler Wirtschafts-Lexikon (1997: 1912).
9   Translated from German.
10 Translated from German.
11 'Architectural innovation is the reconfiguration of an established system to link together existing components in a new way' (Henderson and Clark, 1990: 12).
12 In the term 'low degree of novelty' 'minor innovation', 'routine innovation', 'adopted innovation', 'pretended innovation', 'incremental innovation' and 'cost reduction' shown in Figure 2.6 are summarized.
13 In the terms 'middle degree of novelty', 'architectural innovation', and partly 'ameliorated innovation', as well as 'further development' shown in Figure 2.6. are summarized
14 In the term 'high degree of novelty' 'radical innovation', 'non-routine innovation', 'original innovation', 'new development', 'basic innovation', 'breakthrough innovation' and 'discontinues innovation' shown in Figure 2.6 are summarized.
15 The SBB (Swiss National Railway) passengers should be able to go into the railway car without paying in advance. The passenger would be identified at the entrance or exit of the cars. Data is submitted to a central system that automatically collects the fee from the passengers.
16 The Swiss offer their customers the service of entering the airplane without taking a boarding pass from check-in. The passengers get an electronic boarding pass, which has an electronic key inside. With this key, the check-in is done automatically at passport control. With this system, short connection times are very simple, because no 'material' ticket has to be changed.
17 See Seibert (1998:107); Hauschildt (1993:15); Herzhoff (1991:11); Trommsdorff and Schneider (1990: 3); Kaplaner (1986: 15)
18 Leifer et al. (Leifer, et al. 2000) differentiate between incremental and radical innovations in the dimension of the business case: ' Business case [incremental innovation]: A compete and detailed plan can be developed at the beginning of the process because of the relatively low level of uncertainty. [Radical innovation]: The business model evolves through discovery-based technical and market leanings and likewise the business plan must evolve as uncertainty is reduced.'

19 In contrast to this strategy is the Japanese approach, where they follow the contrary strategy in general, according to Albach, Pay and Okamuro (1991: 309ff.).
20 The author has translated the German word 'dispositiv' used by Hauschildt (1997: 25) as anticipative.
21 Tipotsch (1997: 55) additionally mentions management processes and support processes. But based on the process view (Brockhoff, 1995: 987; Hauschildt, 1993: 23), innovation management is part of the innovation system. Therefore, in this work the management and support processes are added to the innovation and delivery processes.
22 This chapter is partly based on Savioz (2002: 10ff.).
23 'Absorptive capacity' is understood as 'the ability to evaluate and utilize outside knowledge' (Cohen and Levinthal, 1990: 128).
24 For further discussion about knowledge, see Koruna (1999: 43) and Wiegand (1996: 162).
25 See also Probst et al. (1999: 46).
26 In its proper definition, knowledge is created by one individual. However, the organization supports this knowledge creation, which therefore, should be understood as process (Nonaka and Takeuchi, 1995: 74).
27 See Machlup (1980); Nonaka (1991: 96); Krog et al. (1998: 126) and Grant (1996: 163; 2003: 201).
28 Translated from German.
29 Translated from German.
30 Translated from German.
31 Translated from German.
32 According to Altmann (2003), who uses this categorization for defining a product strategy.
33 On p. 22 it was shown that the design, planning and cognitive schools have the most interesting understanding for the strategy definition used in this work. Therefore all other seven schools are not presented in the context of innovation management. Some of these not mentioned concepts can be found in Hoffmann-Ripken (2003: 120ff.) and Burgelman (1983a; 1983b).
34 This concept is restricted for product innovations.
35 Compare also Karlsson and Ahlström (1997: 481).
36 Compare Rechtin and Meier (1997: 21).
37 Compare Hamel and Prahalad (1994: 108).

5 Concept

1 Compare the concept of innovation process of Ropohl (1979: 272).
2 For a more detailed discussion about 'strategic architecture', see p. 50.
3 See p. 35.
4 An opportunity field is an opportunity that has a major impact on a company's strategic goals and path. This term is introduced to make an explicit distinction between opportunities with less strategic impact or major strategic impact.
5 For a more detailed discussion, see Schaad (2001: 104).

6   For a more detailed discussion about the intelligence tools, especially in the context of technology, see Savioz (2002: 62ff.) and Lichtenthaler (2000: 330ff.).
7   For the definition of opportunity field, see p. 88.
8   For a detailed discussion about the definition of functions, innovations fields, business fields, business fields and technology platforms, see pp. 70ff.
9   The process of defining these elements is described more in detail on pp. 70ff.
10 Other creativity methods are presented by Biedermann (2002: 54).
11 For a more detailed discussion about potential creativity barriers, see Biedermann (2002: 52).
12 See also, in the context of innovation fields, pp. 77ff.
13 See also, in the context of functions, pp. 18ff.
14 For a more detailed discussion about scenario technique, see Holt (1988: 139f.).
15 For a detailed discussion about roadmaps, see Bucher (2003).
16 Other calculation methods are, according to Chakravarthy (1986: 440f.) Return on investment, Return on sales, Growth in Revenues, Cash Flow/Investment, Market Share, Market Share Gain, Product Quality Relative to Competitors, New Product Activities Relative to Competitors, Direct Cost Relative to Competitors, Product RandD, Process RandD, Variations in ROI, Percentage Point Change Roi, and Percentage Point Change in Cash Flow/Investment and, according to Edelmann *et al.* (2003: 3), the real option approach is another method.
17 See pp. 118ff.
18 Brodbeck (1999: 99) discusses the make or buy decision in the context of technology. Nevertheless, these findings can also be taken in the context of innovation.
19 Brodbeck (1999: 14) discusses the keep or sell decision in the context of technology. Nevertheless, these findings can also be taken in the context of innovation.
20 A more detailed discussion about 'make or buy/keep or sell' can be found in Brodbeck (1999).
21 For a more detailed discussion about redesigning the organization and processes of the innovation system, compare Schaad (2001).

6 Action Research

1   For a detailed discussion about action research, see p. 8.
2   The key figures are based on the project time.
3   Compare Brodbeck (1999: 50).
4   These three selected criteria were extracted from Bleicher's (1992) concept of integrated management. A company, understood as a system, can be divided into structures, behaviours and objectives on normative, strategic and operational levels. Due to the fact that this work is located at strategic level and does focus on the innovation system of a company, the criteria are restricted to this scope.
5   The criterion, 'innovation decision processes', is selected because of the permanent interaction of the strategic level with the decision process,

which have a tendency to reciprocative influence in terms of a constant reproduction of both (cf. Becker (1996: 143ff.), Giddens (1984: 288ff.) and Brodbeck (1999: 50)).
6 In Figure 6.3, seven strategic technology platforms are visible; one is 'Other technologies'. There are therefore actually only six strategic technology platforms.
7 'Cash Cow' is a term used in the Boston Consulting Matrix. A detailed discussion can be found in Henderson (2003: 42).
8 A detailed explanation about the bullwhip effect is described by Lee and Padmanabhan (1997).
9 The theory for creating an innovation roadmap is described on p. 111.
10 See p. 85 for the background of the different categories.
11 A detailed explanation about the bullwhip effect is described by Lee and Padmanabhan (1997).
12 At MicroSys the opportunity landscape was called opportunity pool.
13 The five forces are: existing competitors, new competitors, substitution competitors, suppliers and customers.
14 Business and technology innovations are integrated directly into the innovation architecture. The organizational innovations are only integrated indirectly because they are derived in the innovation strategy process out of the innovation architecture.

8 Towards a New Set of Management Principles

1 Translated from the German.

Bibliography

Abell, D. F., *Defining the Business. The Starting Point of Strategic Planning* (Englewood Cliffs: Prentice-Hall, 1980).

Abell, D. F., 'Competing Today While Preparing for Tomorrow', *Sloan Management Review*, 40:3 (1999), pp. 73–81.

Abernathy, W. J. and Clark, K. B., 'Innovation: Mapping the Winds of Creative Destruction', *Research Policy*, 14 (1985), pp. 3–22.

Afuah, A., *Innovation Management: Strategies, Implementation, and Profits* (Oxford: Oxford University Press, 1998).

Afuah, A., *Innovation Management: Strategies, Implementation, and Profits* (Oxford: Oxford University Press, 2002).

Albach, H., Pay, D. d. and Okamuro, H., 'Quellen, Zeiten und Kosten von Innovationen, Deutsche Unternehmen im Vergleich zu ihren japanischen und amerikanischen Konkurrenten', *Zeitschrift füer Betriebswirtschaft*, 61 (1991), pp. 309–24.

Allen, T. J., *Managing the Flow of Technology* (Massachusetts Institute of Technology, 1977).

Allen, T. J., *Managing the Flow of Technology* (Massachusetts Institute of Technology, 1986).

Altmann, G., *Unternehmensfüehrung und Innovationserfolg* (Wiesbaden: Deutscher Universitäetsverlag, 2003).

Andrew, J., 'Raising the Return on Innovation – Innovation to Cash Survey 2003', *Boston Consulting Group* (2003).

Andrews, K. R., *The Concept of Corporate Strategy* (Homewood, Illinois: Irwin, 1987).

Ansoff, I. H. and Stewart, J. M., 'Strategies for a Technology-base Business', *Harvard Business Review*, 45:6 (1967), pp. 71–83.

Aregger, K., *Innovation in sozialen Systemen – Einfüehrung in die Innovationstheorie der Organisation* (Bern/Stuttgart, 1976).

Backhaus, K. and Zoeten, R. d., 'Produktentwicklung, Organisation der', in *Handwörterbuch der Organisation*, ed. E. Frese (Stuttgart: Poeschel, 1992), pp. 2024–39.

Baldwin, C. Y. and Clark, K. B., 'Managing in an Age of Modularity', *Harvard Business Review*, September–October (1997), pp. 84–93.

Barnett, H. G., *Innovation: The Basis of Cultural Change* (New York, 1953).

Barron, D., 'Evolutionary Theory', in *The Oxford Handbook of Strategy*, ed. D. O. Faulkner and A. Campell (New York: Oxford University Press, 2003).

Bartley, W., 'Philosophy of Biology versus Philosophy of Physics', in *Evolutionary Epistemology, Theory of Rationality, and Sociology of Knowledge*, ed. G. Radnitzky and W. Bartley (Illinois: Open Court, 1987).

Baum, J. A. C. and Singh, J. V., *Evolutionary Dynamics of Organizations* (Oxford: Oxford University Press, 1994).

Becker, A., *Rationalitäet strategischer Entscheidungsprozesse* (Wiesbaden: Deutscher Universitäetsverlag, 1996).

Becker, S. W. and Whisler, T. L., 'The Innovation Organization. A Selective View of Current Theory and Research', *Journal of Business*, 40 (1967), pp. 462–9.

Beer, S., *Brain of the Firm* (London, 1972).

Beer, S., *The Heart of Enterprise* (London, 1979).

Berth, R., 'Eine Langzeitstudie zeigt, dass deutsche Manager sich nicht an die wirklich renditeträchtigen Neuerungen herantrauen', *Harvard Business Manager*, June (2003), pp. 16–19.

Bessant, J., *High-Involvement Innovation* (New York: John Wileyz & Sons, 2003).

Biedermann, M., 'Course: "Value Engineering Management"' (Zurich, 2002).

Biedermann, M., Tschirky, H., Birkenmeier, B. and Brodbeck, H., 'Value Engineering Management and Handshake Analysis', in *Technologie–Management: Idee und Praxis* (Zürich: Orell Füssli, 1998), pp. 541–52.

Biermann, T., 'Innovation in der Dienstleistung – strategische Optionen', in *Innovation mit System*, ed. T. Biermann and G. Dehr (Berlin/Heidelberg: Springer, 1997), pp. 33–54.

Bleicher, K., *Das Konzept Integriertes Management*, 2nd edn (Frankfurt: Campus, 1992).

Bond, E. U. and Houston, M. B., 'Barriers to Matching New Technologies and Market Opportunities in Established Firms', *Journal of Product Innovation Management*, 20 (2003), pp. 120–35.

Booz, Allen and Hamilton, *New Product Management for the 1980s* (New York, 1982).

Booz, Allen and Hamilton, *Integriertes Technologie- und Innovationsmanagement: Konzepte zur Stärkung der Wettbewerbskraft von High-Tech-Unternehmen*: Berlin: Erich Schmidt, 1991.

Bortz, J. and Döering, N., *Forschungsmethoden und Evaluation* (Berlin: Springer, 1995).

Braun, C.-F., *Der Innovationskrieg: Ziel und Grenzen der industriellen Forschung und Entwicklung* (Müenchen: Hanser, 1994).

Brockhoff, K., 'Innovationsmanagement', in *Handbuch des Marketings* (Stuttgart, 1995), pp. 981–95.

Brockhoff, K. and Chakrabarti, A. K., 'R&D/Marketing Linkage and Innovation Strategy: Some West German Experience', *IEEE Transactions of Engineering Management*, 35:3 (1988), pp. 167–74.

Brodbeck, H., *Strategische Entscheidungen im Technologie-Management* (Züerich: IO, 1999).

Bucher, P., 'Integrated Technology Roadmapping: Design and implementation for technology-based multinational enterprises', Dissertation (Swiss Federal Institute of Technology Zurich: Center for Entwerprise Science, 2003).

Bullinger, H.-J., 'Wegweiser in die Zukunft', in *Kunststüeck Innovation: Praxisbeispiele aus der Fraunhofer-Gesellschaft*, ed. H.-J. Warnecke and H.-J. Bullinger (Berlin/Heidelberg: Springer, 2003).

Bullinger, H.-J. and Auernhammer, K., 'Innovationen im Spannungsfeld von Kreativitäet und Planung', in *Kunststüeck Innovation: Praxisbeispiele aus der Frauenhofer-Gesellschaft*, ed. H.-J. Warnecke and H.-J. Bullinger (Berlin/Heidelberg: Springer, 2003).

Burgelman, R. A., 'Corporate entrepreneurship and strategic management: Insights from a process study', *Management Science*, 29:12 (1983a), pp. 1349–64.

Burgelman, R. A., 'A model of the interaction of strategic behaviour, corporate context, and the concepts of strategy', *Academy of Management Journal*, 8 (1983b), pp. 61–70.
Call, G. and Vöelker, R., 'Innovations-Check', *IO Management*, 5 (1999), pp. 58–63.
Chakravarthy, B. S., 'Measuring Strategic Performance', *Strategic Management Journal*, 7 (1986), pp. 437–58.
Chandler, A. D., *Strategy and Structure. Chapters in the History of Industrial Enterprise* (Cambridge, Mass; London: MIT Press, 1962).
Checkland, P., *System Thinking, System Practice* (Chichester: John Wiley & Sons, 1993).
Cohen, W. M. and Levinthal, D. A., 'Absorptive Capacity: A New perspective on Learning and Innovation', *Administrative Science Quarterly*, 35:1 (1990), pp. 128–52.
Collins, J., *Good to Great* (New York: HarperCollins, 2001).
Cooper, R. G., 'Overall Corporate Strategies for New Product Programs', *Industrial Marketing Management*, 14:3 (1985), pp. 179–94.
Cunningham, J. B., *Action Research and Organizational Development* (London: Praeger Publishers, 1993).
Daenzer, W. F. and Haberfellner, R., *Systems Engineering: Methodik und Praxis* (Zürich: Verlag Industrielle Organisation, 1999).
Damanpour, F., 'Organizational Innovation: A Meta-Analysis of Effects of Determinants and Moderators', *Academy of Management Journal*, 34 (1991), pp. 555–90.
Davis, P., *The American Heritage Dictionary of the English Language* (New York: Dell, 1980).
Davis, S. and Botkin, J., 'The Coming of Knowledge-Based Business', *Harvard Business Review*, 72:5 (1994), pp. 165–70.
Dougherty, D., 'Interpretive barriers to successful product innovation in large firms', *Organization Science*, 3:2 (1992), pp. 179–202.
Drucker, P., 'The Discipline of Innovation', *Harvard Business Review*, May–June (1985), pp. 67–72.
Edelmann, J., Koivuniemi, J., Laaksonen, P. and Sissonen, A., 'Decision making process combined with strategic options approach in innovation proposal selection', R&D Management Conference, Manchester, (2003).
Ehrer, T., *Erfolgreiche Produktinnovation* (Graz, 1994).
Evangelisty, R., Iammarino, S., Mastrostefano, V. and Silvani, A., 'Looking for Regional Systems of Innovation: Evidence from the Italian Innovation Survey', *Regional Studies*, 36:2 (2002), pp. 173–86.
Ewald, A., *Organisation des Strategische Technologie-Managements* (Berlin: Erich Schmidt Verlag, 1989).
Finckh, S., 'Projektarbeit: Innovations-Architektur', in *Center for Enterorise Science* (Zürich: ETH Zürich, 2003).
Flamm, K., *Creating the Computer: Government, Industry and High Technology* (Washington, DC: Bookings Institution, 1988).
Flüehmann, M., 'Identifikation und Bewertung neuer Geschäftsfelder', in *Center for Enterpries Science* (Zürich: ETH Zürich, 2003).
Foster, R. N., *Innovation, The Attacker's Advantage* (New York: Summit Books, 1986).

Foxall, G. and Johnston, B., 'Strategies of Innovation in Smaller Firms', *Technovation*, 6:3 (1987), pp. 169–87.
Gabler, *Wirtschafts-Lexikon – Band 2*, 14th edn (Wiesbaden: Gabler, 1997).
Gäelweiler, A., *Strategische Unternehmensführung*, 2nd edn (Frankfurt: Campus, 1990).
Garud, R. and Kumaraswamy, A., 'Changing Competitive Dynamics in Network Industries: An Exploration of Sun Microsystem's Open Systems Strategy', *Sloan Management Review*, 14 (1993), pp. 351–69.
Garud, R. and Kumaraswamy, A., 'Technological and Organizational Designs for Realizing Economies of Substitution', *Strategic Management Journal*, 16 (1995), pp. 93–109.
Gassmann, O. and Zedwitz, M., *Internationales Innovationsmanagment* (Müenchen: Franz Wahlen, 1996).
Giddens, A., *The Constitution of Society: Outline of the Theory of Structuration* (Cambridge, 1984).
Gilbert, J. T., 'Choosing an innovation strategy: theory and practice', *Business Horizon*, Nov./Dec. (1994), pp. 16–22.
Gordon, J. L., 'Creating Knowledge Maps by Exploiting Dependent Relationships', *Knowledge Based Systems*, 13:April (2000), pp. 71–9.
Grant, R. M., 'Prospering in Dynamically-Competitive Environments: Organizational Capability as Knowledge Integration', *Organizational Science*, 7:4, July–August (1996), pp. 375–87.
Grant, R. M., 'The Knowledge-based View of the Firm', in *The Oxford Handbook of Strategy*, ed. D. O. Faulkner and A. Campel (New York: Oxford University Press, 2003).
Greenwood, D. J. and Levin, M., *Introduction to Action Research. Social Research for Social Change* (Thousand Oaks: Sage Publications, 1998).
Grochla, E., *Einfüehrung in die Organisationstheorie* (Stuttgart: Poechel, 1978).
Haapaniemi, P., 'Innovation: Closing the Implementation Gap', *Chief Executive*, November (2002).
Hamel, G., 'Innovation Now!', *Fast Company*, December:65 (2002).
Hamel, G. and Prahalad, C. K., *Competing for the Future* (Harvard, Mass.: Harvard Business School Press, 1994).
Hannan, M. T. and Freeman, J., 'The Population Ecology of Organizations', *American Journal of Sociology*, 82:March (1977), pp. 929–64.
Hauschildt, J., 'Innovationsmanagement', in *Handwöerterbuch der Organisation*, ed. E. Frese (Stuttgart: Poeschel, 1993), pp. 1029–41.
Hauschildt, J., *Innovationsmanagement* (Müenchen: Franz Vahlen, 1997).
Hayek, F. A., *Die Theorie komplexer Phäenomene* (Tüebingen: Walter Eucken-Institut, 1972).
Hayek, F. A., *Law, Legislation and Liberty* (London, 1973).
Henderson, B. D., 'Das Konzept der Strategie (1980)', in *Das Boston Consulting Group Strategie-Buch*, ed. B.v. Oetinger (Müenchen: Econ, 2003), pp. 26–55.
Henderson, R. M. and Clark, K. B., 'Architectural Innovation: The Reconfiguration of Existing Product Technologies and the Failure of Established Firms', *Administrative Science Quarterly*, 35 (1990), pp. 9–30.
Herzhoff, S., *Innovation-Management, Gestaltung und Prozessen und Systemen zur Entwicklung und Verbesserung der Innovationsfäehigkeit von Unternehmungen* (Bergisch Gladbach: Josef Eul Verlag, 1991).

Hoffmann-Ripken, B. S., *Innovationsstrategie aus einer kognitionstheoretischen Perspektive* (Köeln: Josef Eul Verlag, 2003).
Holt, K., *Product Innovation Management*, 3rd edn (London: Butterworths, 1988).
Horsch, J., *Innovations- und Projektmanagement* (Wiesbaden: Gabler, 2003).
Hunger, J. D. and Wheelen, T. L., *Essentials of Strategic Management*, 3rd edn (Englewood Cliffs, NJ: Prentice-Hall, 2002).
Iansiti, M., *Technology Integration: Making critical choices in dynamic world* (Boston, MA: Harvard Business School Press, 1998).
Kambil, A., 'Good ideas are not enough: Adding execution muscle to innovation engines', *www.accenture.com* (Accenture, March 2003)
Kaplaner, K., *Betriebliche Voraussetzung erfolgreicher Produktinnovationen* (Müenchen: Verlag GBI, 1986).
Karlsson, C. and Ahlströem, P., 'Perspective: Changing Product Development Strategy – a Mangerial Challenge', *Journal of Product Innovation Management*, 5 (1997), pp. 114–28.
Kawai, T., 'Generating innovation through strategic action programmes', *Long Range Planning*, 25:3 (1992), pp. 36–42.
Kieser, A. and Kubicek, H., *Organisation* (Berlin: Walter de Gruyter, 1992).
Knight, K. E., A Descriptive Model of the Intra-Firm Innovation-Process', *Journal of Business*, 40 (1967), pp. 478–96.
Koruna, S., *Kern–Kompetenzen–Dynamik* (Züerich: Orell Füessli Verlag, 1999).
Koruna, S., 'Wissensmanagement', *Internal Report, ETH Züerich* (2001).
Krog, G. V., Venzin, M. and Roos, J., 'Knowledge Management', in *Technologie-Management: Idee und Praxis*, ed. H. Tschirky and S. Koruna (Züerich: Industrielle Organisation, 1998).
Krogh, G. V., Nonaka, I. and Aben, M., 'Making the Most of Your Company's Knowledge: A Strategic Framework', *Long Range Planning*, 34 (2001), pp. 421–39.
Kroy, W., 'Technologiemanagement fuer grundlegende Innovationen', in *Handbuch Technologie-Management*, ed. E. Zahn (Stuttgart: Schäeffer-Poeschel, 1995), pp. 57–79.
Kubicek, H., *Empirische Organisationsforschung: Konzeption und Methodik* (Stuttgart: Poeschel, 1975).
Kuivalainen, O., Kyläeheiko, K., Puumalainen, K. and Saarenketo, S., 'Knowledge-Based View on Internationalization: Finnish Telecom Software Suppliers as an Example', in *Management of Technology: Growth through Business Innovation and Entrepreneurship*, ed. M. Zedtwitz (Amsterdam: Pergamon, 2003).
Lang, H.-C., *Technology Intelligence: Ihre Gestaltung in Abhäengigkeit der Wettbewerbssituation* (Züerich: Industrielle Organisation, 1998a).
Lang, H.-C., *Technology Intelligence: Voraussetzung füer den Technologie Entscheid* (Züerich: Orell Füessli, 1998b).
Lee, H. L. and Padmanabhan, V., 'The Bullwhip Effect in Supply Chains', *Sloan Management Review*, 38:3 (1997).
Leenders, M. A. A. M. and Wierenga, B., 'The effectiveness of different mechanisns for integrating marketing and R&D', *Journal of Product Innovation Management*, 19 (2002), pp. 305–317.
Leifer, R., McDermott, C. M., O'Connor, G. C., Peters, L. S., Rice, M. P. and Veryzer, R. W., *Radical Innovation* (Boston: Harvard Business Review Press, 2000).

Lewin, K., 'Action Research and Minority Problems', *Journal of Social Issues*, 2:4 (1946), pp. 34–46.

Lichtenthaler, E., 'Organisation der Technology Intelligence: Eine empirische Untersuchung in technologieintensiven, international täetigen Grossunternehmen' (Züerich: Dissertation ETH Nb. 13787, 2000).

Lincke, M., *Diplomarbeit: Erfassung und Bewertung von Technologien* (ETH Zürich: Center for Enterprise Science, 2004).

Machlup, F. M., *Knowledge: Its Creation Distribution and Economic Significance* (Princeton: Princeton University Press, 1980).

Malik, F., *Strategie des Managements komplexer Systeme* (Bern/Stuttgart, 1992).

Malik, F., *Management – Perspektiven-Wirtschaft und Gesellschaft, Strategie, Management und Ausbildung*, 3rd edn (Bern/Stuttgart: Haupt, 2001a).

Malik, F., 'Strategische Unternehmungsführung als Steuerung eines komplexen Systems', in *MZSG Seminardokumentation* (St Gallen, 2001b).

Malik, F., *Strategie des Managements komplexer Systeme*, 7th edn (Bern/Stuttgart: Haupt, 2002).

Marquis, D. G., *Successful Industrial Innovations: A Study of Factors Underlying Innovation in Selected Firms* (Washington, DC: National Science Foundation, 1969).

Martensen, A. and Dahlgaard, J., 'Integrating Business Excellence and Innovation Management: Developing Vision, Blueprint and Strategy for Innovation in Creative and Learning Organizations', *Total Quality Management*, 10:4/5 (1999), pp. 627–35.

Maurer, P. J., *Das Informationsmanagement der Innovatik* (Züerich, 2002).

Meier, M., 'Produktfunktion festlegen' (EC E-Collection, July 2003) internet source of 2.7.2003: http://e-collcetion.ethbib.ethz.ch/cgi-bin/show.pl?type=Lehr&nr=40.

Meier, M., Fadel, G., Wälchi, N., Kobe, C. and John, C. S. 'The Impact Model for Innovation Success and its Assessment', (2004).

Mensch, G., *Das technologische Patt – Innovation üeberwindet die Depression* (Frankfurt am Main: Umschau Verlag, 1975).

Mikkola, J. H. and Gassmann, O., 'Modeling Modularity of Product Architectures', in *Management of Technology: Growth through Business Innovation and Entrepreneurship*, ed. M. Zedtwitz (Amsterdam: Pergamon, 2003).

Mintzberg, H. and Lampel, J., 'Reflecting on the Strategy Process', *Sloan Management Review*, Spring (1999), pp. 21–30.

Müeller-Stewens, G. and Lechner, C., *Strategisches Management* (Stuttgart: Schäeffer-Poeschel, 2001).

Nadler, D. A., 'Organisations-Architektur', in *Organisations-Architektur*, ed. Campus (Frankfurt/New York, 1994), pp. 12–20.

Nelson, R. and Winter, S., *An Evolutionary Theory of Economic Change* (Cambridge: Belknap, 1982).

Nonaka, I., 'The Knowledge-Creating Company', *Harvard Business Review*, November–December (1991), pp. 96–104.

Nonaka, I. and Takeuchi, H., *The Knowledge-Creating Company: How Japanes Companies Create the Dynamics of Innovation* (New York: Oxford University Press, 1995).

Olschowy, W., *Externe Einflussfaktoren im Strategischen Innovationsmanagement* (Berlin: Erich Schmidt Verlag, 1990).

Orten, J. D. and Weick, K. E., 'Loosely coupled systems: a re-conceptualization', *Academy of Management Review*, 15: pp. 103–223 (1990).

Pahl, G. and Beitz, W., *Konstruktionslehre: Methoden und Anwendung* (Berlin: Springer, 1993).
Pavitt, K., *Technical Innovation and British Economic Performance* (London: MacMillan Press, 1980).
Peiffer, S., *Technologie-Früehaufkläerung: Identifikation und Bewertung zuküenftiger Technologien in der strategischen Unternehmensplanung* (Hamburg: S&W Verlag, 1992).
Pfeiffer, W., 'Strategisches Technologie-Management bei anlageintensiven Produktionstrukturen', *Kongresses Anlagenwirtschaft*, Nüernberg (1991).
Pfeiffer, W., Weiss, E., Volz, T. and Wettengl, S., *Funktionalmarkt-Konzept zum strategischen Management prinzipieller technologischer Innovationen* (Göettingen: Vandenhoeck und Ruprecht, 1997).
Pherson, J. M. and Ranger-Moore, F. R., 'Evolution on a Dancing Landscape: Organizations and Networks in Dynamic Blau Space', *Social Forces*, 70 (1994), pp. 19–42.
Picot, A., Reichwald, R. and Nippa, M., 'Zur Bedeutung der Entwicklungsaufgabe füer die Entwicklungszeit – Ansäetze füer die Entwicklungszeitgestaltung', *ZfbF-Sonderheft*, 23 (1988), pp. 112–37.
Polanyi, M., *The Tacit Dimension* (London: Routledge & Kegan Paul, 1966).
Popp, R., 'Methodik der Handlungsforschung', in *Wie kommt Wissenschaft zu Wissen? Einfüehrung in die Forschungsmethodik und Forschungspraxis*, ed. T. Hug (Schneider verlag Hohengehren GmbH, 2001), pp. 400–12.
Porter, M. E., *Competitive Strategy: Techniques for Analyzing Industries and Competitors* (New York: Free Press, 1980).
Porter, M. E., *Competitive Advantage: Creating and Sustaining Superior Performance* (New York: Macmillan, 1985).
Porter, M. E., *The Competitive Advantage of Nations* (New York: Free Press, 1990).
Porter, M. E., 'What is Strategy?', *Harvard Business Review*, 74:6 (1996), pp. 61–78.
Prahalad, C. K. and Hamel, G., 'The Core Competence of the Corporation', *Harvard Business Review*, May–June (1990), pp. 79–91.
Probst, G., Raub, S. and Romhardt, K., *Wissen managen: Wie Unternehmen Ihre wertvollste Ressource optimal nutzen* (Wiesbaden: Gabler, 1999).
Quinn, J. B., *Strategies for Change: Logical Incrementalism* (Homewood, Illinois: Irwin, 1980).
Quinn, J. B., 'Managing Innovation: Controlled Chaos', *Harvard Business Review*, May–June (1985), pp. 73–84.
Rammert, W., *Das Innovationsdilema – Technikentwicklung in Unternehmungen* (Opladen: Westdeutscher Verlag, 1988).
Rappaport, A., 'Creating Shareholder Value: The New Standard for Business Performance', (New York: Free Press, 1986).
Rechtin, E. and Meier, M., *The Art of Systems Architecting* (New York: CRC Press, 1997).
Ropohl, G., *Eine Systemtheorie der Technik* (Müenchen: Hanser, 1979).
Rosenberg, N., 'Innovation's uncertain terrain', *McKinsey Quartlery*, 3 (1995), pp. 171–85.
Sanchez, R., 'Managing Knowledge into Competence: The Five Learning Cycles of the Competent Organization', in *Knowledge Management and Organizational Competence*, ed. R. Sanchez (Oxford; New York: Oxford University Press, 2001), pp. 3–37.

Sanchez, R. and Mahoney, J. T., 'Modularity, Flexibility, and Knowledge Management in Product and Organisation Design', *Strategic Management Journal*, 17 (1996), pp. 63–76.

Savioz, P., 'Technology Intelligence in Technology-based SMEs in Technology- and Innovation Management' (Züerich: Dissertation ETH Züerich 2002).

Savioz, P., 'Competence Management with the Opportunity Landscape', in *Technology and Innovation Management on the Move – From Managing Technology to Managing Innovation-driven Enterprises*, ed. P. Savioz (Zürich: Orell Füssli, 2003), pp. 191–200.

Savioz, P., *Technology Intelligence* (Palgrave Macmillan, 2004).

Schaad, D., *Modellierung unternehmensspezifischer Innovations-Prozessmodelle* (Züerich: ETH Züerich, 2001).

Schendel, D. E. and Cool, K., 'Development of the Strategic Management Field: Some Accomplishments and Challenges', in *Strategic Management Frontiers*, ed. J. H. Grant (Greenwich: JAI Press, 1988), pp. 17–33.

Schlaak, T. M., *Der Innovationsgrad als Schlüesselvariable, Perspektiven füer das Management von Produktentwicklungen* (Wiesbaden: Deutscher Universitäets-Verlag GmbH, 1999).

Schlegelmilch, G., *Management strategischer Innovationsfelder: Prozeßbasierte Integration markt- und technologieorientierter Instrumente* (Wiesbaden: Gabler, 1999).

Schofield, B. A. and Feltmate, B. W., 'Sustainable Development Investing', *Employee Benefits Journal*, 28:1 (2003), pp. 17–21.

Schumpeter, J. A., 'Unternehmer', in *HW der Staatswissenschaften*, ed. L. Elster and A. Wieser (Jena, 1927), pp. 476–87.

Schumpeter, J. A., *Business Cycle: A Theoretical, Historical and Statistical Analysis of Capitalist Process* (New York: McGraw-Hill, 1939).

Seibert, S., *Technisches Management: Innovationsmanagement, Projektmanagement und Qualitaetsmanagement* (Leipzig: Teubner, 1998).

Servatius, H.-G., *Methodik des strategischen Technologie-Managements* (Wiesbaden, 1985).

Sherman, P. M., *Strategic Planning for Technology Industries* (Boston, 1982).

Silverstein, M. J., 'Schachspiel und Geschäftsleben (1986)', in *Das Boston Consulting Group Strategie-Buch*, ed. B.v. Oetinger (München: Econ, 2003), pp. 11–14.

Snyder, R., Raben, C. and Farr, J., 'A Model for the Systemic Evaluation of Human Resource Development Programs', *Academy of Management Review*, 5:3 (1980), pp. 431–44.

Sommerlatte, T., '1000 Unternehmen antworten: Die Innovationswelle kommt', in *Arthur D. Little International: Management der Geschäfte von Morgen* (Wiesbaden: Gabler, 1987a), pp. 1–16.

Sommerlatte, T., Layng, B. L. and Oene, F., 'Innovationsmanagement: Schaffung einer inovativen Unternehmenskultur', in *Management der Geschäfte von Morgen*, International ed. Arthur D. Little International (Wiesbaden: Gabler, 1987b), pp. 55–74.

Souder, W. E., 'Managing Relations Between R&D and Marketing in New Product Development Projects', in *The Human Side of Managing: Technology Innovation*, ed. R. Katz (New York, Oxford: Oxford University Press, 2004).

Sowa, J. F., *Conceptual Stuctures: Information Processing in Mind and Machine* (Reating, MA: Addison-Wesley, 1984).

Stahl, H. K. and Eichen, S. F. V. D., 'Vorsicht, "Innovationsmanagement"', *New Management*, 5 (2003), pp. 14–22.

Teece, D. J., 'Contributions and Impediments of Economic Analysis to the Study of Strategic Management', in *Perspectives on Strategic Management*, ed. J.W. Fredrickson (New York: Harper Business, 1990), pp. 39–80.

Teece, D. J., Pisano, G. and Shuen, A., 'Dynamic Capabilities and Strategic Management', *Strategic Management Journal*, 18:7 (1997), pp. 509–33.

Thom, N., *Zur effizienz betrieblicher Innovationsprozesse* (Koenigstein: Peter Hanstein Verlag, 1976).

Thom, N., *Grundlagen des betrieblichen Innovationsmanagements* (Koenigstein: Peter Hanstein Verlag, 1980).

Tidd, J., Bessant, J. and Pavitt, K., *Managing Innovation*, 2nd edn (Chichester: John Wiley & Sons, 2001).

Tipotsch, C., 'Business Modelling: Vorgehensmethodik und Gestaltungsmodelle' (Graz: Technische Universitäet Graz, 1997).

Trommsdorff, V. and Schneider, P., 'Grundzüege des betriebswirtschaftlichen Innovationsmanagements', in *Innovationsmanagement in kleinen und mittleren Unternehmen*, ed. V. Trommsdorf (Müenchen: Franz Vahlen, 1990), pp. 1–25.

Tschirky, H., 'Konzept und Aufgaben des Technologie-Managements', in *Technologie-Management: Idee und Praxis* (Züerich: Orell Füessli, 1998), pp. 193–394.

Tschirky, H., 'The Concept of the Integrated Technology and Innovation Management', in *Technology and Innovation Management on the Move – From Managing Technology to Managing Innnovation-driven Enterprises*, ed. H. Tschirky, H.-H. Jung and P. Savioz (Züerich: Orell Füessli, 2003a), pp. 43–106.

Tschirky, H., 'The Concept of the Integrated Technology and Innovation Management', in *Technology and Innovation Management on the Move – From Managing Technology to Managing Innnovation-driven Enterprises*, ed. P. Savioz (Züerich: Orell Füessli, 2003b), pp. 40–106.

Tushman, M. L. and Anderson, P., *Managing Strategic Innovation and Change* (New York: Oxford University Press, 1997).

Tushman, M. L., Anderson, P. C. and O'Reilly, C., 'Technology Cycles, Innovation Streams, and ambidextrous Organizations: Organization Renewal through Innovation Streams and Strategic Change', in *Managing Strategic Innovation and Change* (New York: Oxford University Press, 1997), pp. 3–23.

Ulrich, H. and Probst, G. J. B., *Anleitung zum ganzheitlichen Denken und Handeln: Ein Brevier füer Füehrungskräefte* (Bern, 1988).

Urabe, K., 'Innovation under Japanese Management', in *Innovation and Management*, ed. K. Urabe, J. Child and T. Kagono (Berlin: Walter de Gruyter, 1988), pp. 3–26.

Utterback, J. M., *Mastering the Dynamics of Innovation* (Boston, MA: Harvard Business School Press, 1994).

Van Maanen, J., 'The fact of fiction in organization ethnography', *Administrative Science Quarterly*, 24:4 (1979), pp. 539–50.

Vahs, D. and Burmester, R., *Innovationsmanagement: Von der Produktidee zur erfolgreichen Vermarktung* (Stuttgart: Schäeffer-Poeschel Verlag, 1999).

VDI, 'Funktionsanalyse: Grundlagen und Methode', *VDI 2803* (1996).

Wagner, K. H., 'Darstellung von Wissen', internet source of 8.12.2003: http://www.fb10.unibremen.de/linguistik/khwagner/semantik/wissen.htm.

Wagner, M. and Kreuter, A., 'Erfolgsfaktoren innovativer Unternehmen', *IO Management*, 10 (1998), pp. 34–41.

Ward, V., 'Metting Meta Knowledge', in *Knowledge Management Review* (1998), pp. 10–15.

Warnecke, H.-J., 'Innovation in Technik und Gesellschaft – Notwendigkeit und Hemmnisse', in *Kunsstüeck Innovation: Praxisbeispiele aus der Frauenhofer-Gesellschaft*, ed. H.-J. Warnecke and H.-J. Bullinger (Berlin/Heidelberg: Springer, 2003).

Welge, M. K. and Al-Laham, A., 'Strategisches Management, Organisation', in *Handwöerterbuch der Organisation*, ed. E. Frese (Stuttgart: Poeschel, 1992), pp. 2355–74.

Wheelwright, S. C. and Clark, K. B., 'Creating Project Plans to Focus', *Harvard Business Review*, March–April (1992), pp. 70–82.

Whitehead, A. N., 'Representation Models and System Architecting', in *The Art of Systems Architecting*, ed. E. Rechtin and M. W. Maier (Boca Raton, Florida: CRC Press, 1997), pp. 118–40.

Wiegand, M., *Prozesse organisationlen Lernens* (Wiesbaden: Gabler, 1996).

Witte, E., 'Innovationsfäehige Organisation', *Zeitschrift füer Organisation*, 42 (1973), pp. 17–24.

Wollmann, H. and Gerd-Michael, H., 'Sozialwissenschaftliche Untersuchungsregeln und Wirkungsforschung', in *Res. Publica. Studium zum Verfassungswesen. Dolf Sternberger zum 70. Geburtstag*, ed. P. Haugs (Müenchen: Fink, 1977).

Wright, R., Pringle, C. and Kroll, M., *Strategic Management: Text and Cases* (Needham Heights: Allyn and Bacon, 1992).

Yin, R. K., *Case Study Research: Design and Methods*, 2nd edn (New York: Sage Publications Inc., 1994).

Zahn, E., 'Innovations- und Technologiemanagement: Eine strategische Schlüesselfrage der Unternehmen', in *Technologie- und Innovationsmanagement*, ed. E. Zahn (Berlin, 1986), pp. 9–48.

Zahn, E., 'Gegenstand und Zweck des Technologiemanagements', in *Handbuch Technologiemanagement*, ed. E. Zahn (Stuttgart: Schäeffer-Poeschel Verlag, 1995), pp. 3–32.

Zahn, E. and Weidler, A., 'Verwertung technologischer Fähigkeiten', in *Handbuch Technologiemanagement*, ed. E. Zahn (Stuttgart: Schäeffer-Poeschel Verlag, 1995), pp. 351–76.

Zöergiebel, W. W., *Technologie in der Wettbewerbsstrategie* (Berlin, 1983).

Index

action research 8, 130
architectural blueprint 204
architecture 50
   strategic 51

case study 7

delivery system 35
design 14
develop 14
direct 14

function 74, 206
   level of abstraction 75
   level of detail 76

innovation 1, 24
   architecture 5, 67
   barriers 26
   field 78
   integrated 33
   management 34
   newness 28
   object 32
   organizational 33
   process 35, 121
   product 33
   roadmap 111
   social 35
   strategy 38
   strategy formulation 42
   system 35
   technological 33
innovativeness 26

keep or sell 116
key buying factors 103

key consuming factors 103
key success factors 102
knowledge 36
   meta 83
   methodological 82
   object 71
   scientific 74
make or buy 116
management 13
   normative 88
   operational 90
   strategic 88
module 72

net present value 111

opportunity landscape 91

products 72

scenario technique 104
strategic fit 101
strategy 18
   formulation 21
system 16
   complexity 16
   evolution 17
   model 48
   systemic interaction 17

technology 73
   attractiveness 114
   portfolio 114
   process 73
   product 73
   strength 114